KU-688-598

RSPB

What's that
FLOWER?

David Burnie

DK

LONDON, NEW YORK, MUNICH, MELBOURNE, AND DELHI

DK LONDON
Project Art Editor Francis Wong
Senior Editor Angeles Gavira
Editorial Assistant Lili Bryant
Pre-production Producer
Nikoleta Parasaki
Producer Alice Sykes
Jacket Designer Mark Cavanagh
CTS Sonia Charbonnier
Managing Art Editor Michelle Baxter
Publisher Sarah Larter
Art Director Philip Ormerod
Associate Publishing Director
Liz Wheeler
Publishing Director Jonathan Metcalf

DK DELHI
Deputy Managing Art Editor
Sudakshina Basu
Design Consultant Shefali Upadhyay
Managing Editor Rohan Sinha
Senior Art Editor Anuj Sharma
Senior Editor Vineetha Mokkil
Designers Kanika Mittal,
Divya PR, Upasana Sharma
Editor Rupa Rao
DTP Designer Shanker Prasad
DTP Manager/CTS Balwant Singh
Production Manager Pankaj Sharma
Picture Researcher Aditya Katyal

First published in 2013 by
Dorling Kindersley Limited
80 Strand, London WC2R 0RL

Penguin Group (UK)

2 4 6 8 10 9 7 5 3 1
001 – 192169 – Mar/2013

Copyright © 2013
Dorling Kindersley Limited

A CIP catalogue record for this book
is available from the British Library
ISBN 978-1-40932-441-6

Printed and bound in China by
South China Printing Co. Ltd.

Discover more at
www.dk.com

ABOUT THE AUTHOR

David Burnie is a Zoology graduate and Fellow of the Zoological Society of London. Having started out as a nature-preserve ranger and biologist, he has since forged a successful career as a writer and consultant specializing in wildlife and plants. He has written and contributed to more than 140 books and multimedia titles on nature and the environment, including DK's *Animal* and *Natural History* encyclopedias. He was shortlisted for the Royal Society's Aventis Prize in 2005. He currently lives and works in France.

Contents

Introduction

This book will help you explore one of the riches of the
European countryside – its wonderful heritage of wildflowers.
In all kinds of habitats, from hedges and meadows to
pathways and pavements, a fascinating sequence of flowers
unfolds each year. Big, small, and amazingly diverse, they all
carry out the same important job: to exchange pollen and
produce seeds, so that plants can reproduce.

    Many wildflower guides are organized according to plant
families, but this book uses a simpler and easier approach: in
the catalogue section (pp.18–103), flowers are arranged by
colour and then by size, helping you to easily home in on the
particular flower that you want to identify. The flower gallery
(pp.104–117) shows you how flowers fit into plant families.
There are thousands of wildflowers in Europe, and no field
guide can cover them all. But if you start by spotting the
commonest, your knowledge will grow fast. Before long, you
will be able to recognize many beautiful and remarkable
plants that add colour and variety to the great outdoors.

David Burnie

Plant Anatomy

To identify plants, you need a basic working knowledge of their parts – particularly flowers and leaves. Usually, both will be present at flowering time, although a small number of plants bloom before their leaves appear. Fruit and seeds (p.14) are also important clues for identification.

Sepal

Leaf margin

Petal

Flower stalk

Leaf stalk

Bract

Leaf

SWEET VIOLET

Stems, leaves, and flowers

Instead of growing at random, leaves and flowers sprout from points called nodes, which are arranged at fixed intervals along plant stems. In some plants, such as Yellow Archangel, flowers are attached directly to the stems. In others, such as Sweet Violet, the flowers grow singly on slender stalks.

Internode

Node

YELLOW ARCHANGEL

Shape and Growth

Plants have characteristic shapes and ways of growing, which can help you to identify these even when they are not in bloom. Some form patches or mats, while others grow singly or in scattered groups. Climbers cling to other plants, using these for support as they grow towards the light.

AQUATIC
Waterlilies root in mud at the bottom of ponds and lakes. Their leaves and flowers grow on the water surface.

CREEPING
Creeping Buttercup spreads by long, overground stems called runners. These runners produce new plants at their nodes.

PATCH-FORMING
Wood Sorrel spreads through underground stems. Taller plants, such as nettles, often spread like this, forming dense clumps.

COLONIAL
Primroses typically grow in scattered groups, or colonies. These can contain a few dozen plants or many thousands.

CLIMBING
Climbers hang on by tendrils, hooks, or by twining their way around solid supports. Bittersweet clambers through other plants.

UPRIGHT
Foxgloves have tall, upright stems and flowerheads containing dozens of blooms. The tall flowerheads are easy for insects to spot.

Flower Anatomy

You can identify most flowers easily with the naked eye. The number of petals is often a key feature, as well as their colour and shape. Most flowers contain male and female reproductive organs. The male stamens and female stigmas sometimes protrude from the flower.

Anther at tip of stamen

Petal

Stigma

Sepal

SCARLET PIMPERNEL

Flower Types

The flowers of Scarlet Pimpernel and Wood Sorrel have five petals and five sepals – the flaps that protect the flower as a bud. Michaelmas Daisy and Slender Thistle flowers are made of many florets – a characteristic feature of the daisy family (pp.36–37).

Sepal

Flower stalk

WOOD SORREL

Disc florets

Ray florets

Short, spiny bracts

Ray florets

MICHAELMAS DAISY

SLENDER THISTLE

Flower Shapes

A flower's petals can be separate or they may fuse to form a tube. The simplest flowers are round, but some plants – such as peas and orchids – have complex flowers that are easy to recognize.

Petals form a tube

Separate petals

COMMON MALLOW

FOXGLOVE

Lateral petal or wing

Upper petal or standard

Pink sepal

COMMON RESTHARROW

Prominent lip

BEE ORCHID

Flower Arrangement

Many plants have solitary flowers, with each one growing on a separate stalk. At the other extreme, some of the most eye-catching plants have flowers in large clusters.

Solitary flowers

Deep violet-purple sepal

Fleshy, oval petal

Rounded petal

Toothed petal

WHITE WATERLILY

WELSH POPPY

MAIDEN PINK

PASQUE-FLOWER

Flowers in clusters

Flowerhead

Umbrella-shaped cluster

Upright column

One-sided flower cluster

DANDELION

HOGWEED

BLUEBELL

AGRIMONY

Leaves

One of the best ways to identify plants is to look at their leaves. Most plants have one type of leaf. In some, however, the lower leaves differ in size and shape from the ones higher up on the stems. As well as noting shape, look carefully to see how the leaves are arranged.

Glossy surface

Veins

Midrib (central vein)

Leaf anatomy
While leaves typically have a stalk, some are attached directly to the stem. Many leaves have a visible network of spreading or parallel veins, and an often-prominent midrib. They can also have winding tendrils, which the plant uses to climb.

Leaf stalk

LESSER CELANDINE

Leaf Shapes

Leaves are extremely varied, but most kinds have a single blade, with or without teeth around its edge. Divided leaves are split into several leaflets or deep lobes, attached to the same leaf stalk.

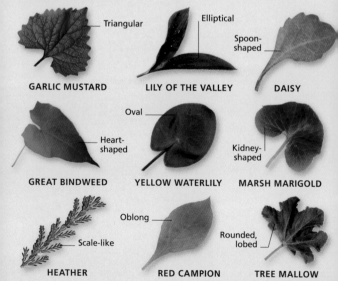

Triangular

GARLIC MUSTARD

Elliptical

LILY OF THE VALLEY

Spoon-shaped

DAISY

Heart-shaped

GREAT BINDWEED

Oval

YELLOW WATERLILY

Kidney-shaped

MARSH MARIGOLD

Scale-like

HEATHER

Oblong

RED CAMPION

Rounded, lobed

TREE MALLOW

Leaf Arrangement

The way that leaves are arranged is as important as their shape. Alternate leaves grow singly along the stem. Opposite leaves are arranged in pairs, with neighbouring pairs usually at right angles to each other. Whorled leaves grow in rings.

Alternate

Opposite

Whorls

BILBERRY

GREATER STITCHWORT

CLEAVERS

Barrel-shaped

Lance-shaped

Arrow-shaped

WHITE STONECROP

SHEEP'S SORREL

VIPER'S BUGLOSS

Three-lobed

Three-part

Deeply lobed

YELLOW CORYDALIS

HERB BENNET

WOOD SORREL

Wedge-shaped

Ladder-like

Hand-like

CREEPING BUTTERCUP

CREEPING CINQUEFOIL

KIDNEY VETCH

Fruit and Seeds

Long after their flowers have withered, many plants can be recognized by their fruit. These can be soft and juicy, but others – such as capsules and pods – are hard and dry.

Feathery seed

TRAVELLER'S JOY

Feathery "clock"

DANDELION

Hooked seed

HERB BENNET

Seed capsule

RED CAMPION

Seed pod

BROAD-LEAVED EVERLASTING PEA

Berry

MISTLETOE

Plant Habitats

Most plants grow in particular habitats, which helps in identifying the flowers that you find. Some are less fussy – they thrive along paths and roadsides, or in places where the ground has been disturbed. These plants include many common weeds.

GRASSLANDS
Clovers and orchids are often a sign of old meadows and pastures.

WOODLAND
Like many woodland plants, Wood Anemone flowers in early spring.

HEATHLANDS
Heather thrives in open heathland, where it grows on acidic soil.

WETLANDS
Yellow Flag is a typical plant of ponds and slow-flowing water.

MOUNTAINS
Yellow Mountain Saxifrage is a typical low-growing upland plant.

COASTS
Many coastal plants, including Sea Rocket, have fleshy leaves.

FLOWER PROFILES

The following pages will help you to identify more than
150 of Europe's most common wildflowers. The profiles
are organized first by colour, and then by size. When
you are looking at a plant, remember that the same
species can often vary – flower sizes, in particular, are
only a general guide.

Flower sizes

Small Less than 1 cm

Medium 1–2 cm

Large More than 2 cm

Symbols

↕ Height

↔ Width

1 GREEN FLOWERS

Even though they may have a long blooming period, green flowers are often small and therefore unobtrusive. However, instead of growing singly, many are arranged in distinctive clusters or flowerheads, which makes them easier to spot.

Small Flowers

Many plants have small green flowers, which are pollinated by insects or by the wind. They include some common weeds.

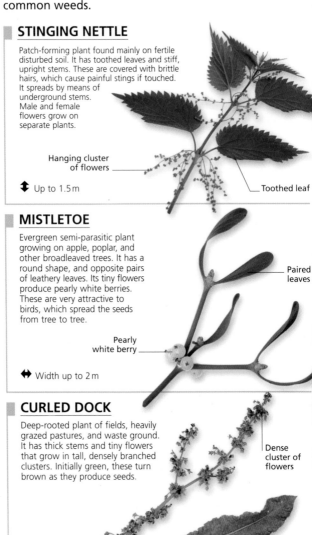

STINGING NETTLE

Patch-forming plant found mainly on fertile disturbed soil. It has toothed leaves and stiff, upright stems. These are covered with brittle hairs, which cause painful stings if touched. It spreads by means of underground stems. Male and female flowers grow on separate plants.

Hanging cluster of flowers

Toothed leaf

✚ Up to 1.5 m

MISTLETOE

Evergreen semi-parasitic plant growing on apple, poplar, and other broadleaved trees. It has a round shape, and opposite pairs of leathery leaves. Its tiny flowers produce pearly white berries. These are very attractive to birds, which spread the seeds from tree to tree.

Paired leaves

Pearly white berry

↔ Width up to 2 m

CURLED DOCK

Deep-rooted plant of fields, heavily grazed pastures, and waste ground. It has thick stems and tiny flowers that grow in tall, densely branched clusters. Initially green, these turn brown as they produce seeds.

Dense cluster of flowers

Large, curly-edged leaf

✚ Up to 1.5 m

SEA BEET

Wild ancestor of beetroot and sugar beet, found on coastal shingle and dry, salty ground close to the shore. It has sprawling stems, and tiny flowers that grow in small clusters. The whole plant sometimes has a reddish tinge.

Small flower cluster on slender stem

Slightly fleshy, glossy leaf

✚ Up to 1 m

FAT HEN

Widespread weed of fields and farmyards; grows rapidly on fertile soils. It has spreading or upright stems and alternate, narrow or diamond-shaped grey-green leaves. Its stems and leaves have a powdery feel.

Ridged stem

Cluster of tiny flowers

✚ Up to 1.2 m

GLASSWORT

Succulent, low-growing plant of mudflats and salt marshes, where its upright stems often carpet the ground. The stems have numerous joints, and bear pairs of tiny leaves and minute flowers. Young plants are bright green, but they often turn red with age.

Scale-like leaf

Plump, jointed stem

✚ Up to 30 cm

»

NAVELWORT

Fleshy, clump-forming plant of old walls and rocky places. From midsummer onwards, the plant produces one or more tall stems with densely packed, bell-shaped flowers, varying in colour from green to pinkish red.

Flowering stem

Dimpled, coin-shaped leaf

↕ Up to 40 cm

PETTY SPURGE

Bright green weed of farmland and gardens, which produces poisonous white sap if cut. Its main stem typically branches into three, creating a flat-topped flowerhead with tiny flowers. It has small, alternate leaves.

Leaf-like bracts

Green-yellow flower

Lobed seed capsule

↕ Up to 40 cm

RIBWORT PLANTAIN

Widespread plant of fields, meadows, and waysides, with a rosette of tapering leaves. From spring to late summer, it produces wiry flowering stems, each topped by a dark-coloured flowerhead. The flowers are tiny, with prominent yellow anthers.

Greenish brown flower

Leafless flowering stem

Parallel veins on leaf

↕ Up to 50 cm

Medium-sized Flowers

Stinking Hellebore is one of the few plants with medium-sized green flowers. It blooms very early in the year.

STINKING HELLEBORE

Flowering early in spring, this robust woodland plant has deeply lobed leaves and a distinctive, foul smell. Its pale yellowish green flowers are small and grow in hanging clusters. The whole plant is poisonous.

Prominent sepal with purple rim

Bell-shaped flower

Hand-like leaf

✦ Up to 80 cm

Large Flowers

Large green flowers are rare. One common plant – Lords and Ladies – has a large green hood, but its individual flowers are small.

LORDS AND LADIES

This shade-loving plant, also called Cuckoo Pint, is unmistakable when in bloom. Its tiny flowers grow at the bottom of a club-shaped flowerhead, which is enveloped by a large hood, or spathe. The flowers produce poisonous berries that are red when ripe.

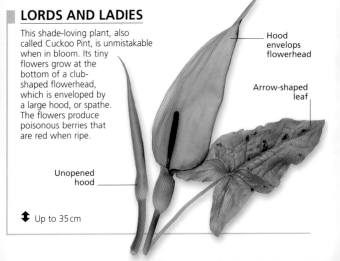

Hood envelops flowerhead

Arrow-shaped leaf

Unopened hood

✦ Up to 35 cm

Carrot Family

With over 2,500 species worldwide, the carrot family contains many wildflowers and aromatic herbs. As a group, they are easy to recognize because most have umbrella-shaped flowerheads.

Plants in this family are often tall and upright, with hollow and sometimes bristly stems. Their flowerheads – called umbels – are usually flat-topped or domed. A single flowerhead can have dozens of spokes and hundreds of tightly packed flowers.

Spring show
Cow Parsley (above) is a typical umbellifer, or member of the carrot family. It sprouts from ground level each spring, and has multiple flowerheads. The flowers are pollinated by a wide range of insects, from hoverflies to bees.

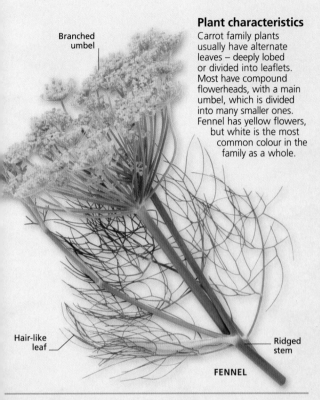

Plant characteristics
Carrot family plants usually have alternate leaves – deeply lobed or divided into leaflets. Most have compound flowerheads, with a main umbel, which is divided into many smaller ones. Fennel has yellow flowers, but white is the most common colour in the family as a whole.

Branched umbel

Hair-like leaf

Ridged stem

FENNEL

Flower anatomy
Carrot family plants usually have five-petalled flowers, although in some, the petals are so small that they can be difficult to see. Once the flowers have been pollinated, they produce dry fruit containing two seeds. The fruit's shape varies, and can be a useful clue in identifying a plant.

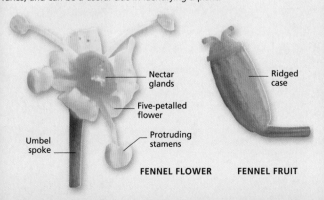

Nectar glands

Five-petalled flower

Umbel spoke

Protruding stamens

Ridged case

FENNEL FLOWER

FENNEL FRUIT

2 WHITE FLOWERS

White is one of the commonest colours among wildflowers. The earliest to bloom is often the Snowdrop, followed by a host of other species – particularly in the cress and carrot familes – as spring and summer get underway.

Small Flowers

Plants with small, white flowers often have them clustered in large groups. They include many species in the cress and carrot families.

GARLIC MUSTARD

Tall, spring-flowering plant of bare ground and hedgerows, often found on lime-rich soil. Its stems are topped by clusters of small flowers. The plant smells of garlic when crushed.

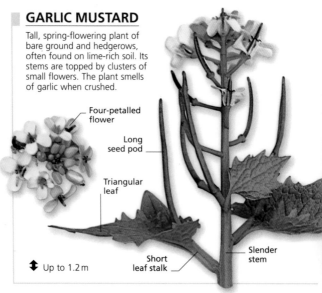

Four-petalled flower

Long seed pod

Triangular leaf

Short leaf stalk

Slender stem

✦ Up to 1.2 m

SWEET ALISON

Low-growing plant typically found on sandy and rocky coasts, and on bare ground inland. Its sweet-smelling flowers are clustered at the tips of the stems, which lengthen as the plant produces seeds.

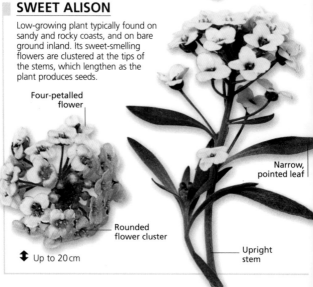

Four-petalled flower

Narrow, pointed leaf

Rounded flower cluster

Upright stem

✦ Up to 20 cm

SHEPHERD'S PURSE

Widespread weed found in many habitats, from fields and gardens to cracks in paving, where it is often just a few centimetres high. Its unobtrusive flowers produce seed cases that resemble old-fashioned purses. The ground-hugging leaves form a rosette.

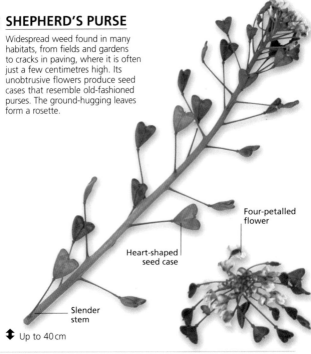

Four-petalled flower

Heart-shaped seed case

Slender stem

⬍ Up to 40 cm

WATERCRESS

Sprawling, succulent plant of wet places and shallow streams. It has edible, peppery-tasting stems and leaves. Its leaves have dark green, glossy leaflets and its flowers grow in dense clusters that elongate as they produce seeds.

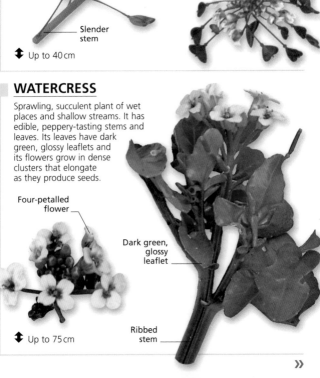

Four-petalled flower

Dark green, glossy leaflet

Ribbed stem

⬍ Up to 75 cm

»

WHITE STONECROP

Fleshy, mat-forming plant of dry places, from rocks to gravel paths. The waxy leaves are often tinged with red, and the stems develop roots as they spread. Its star-shaped flowers have five petals.

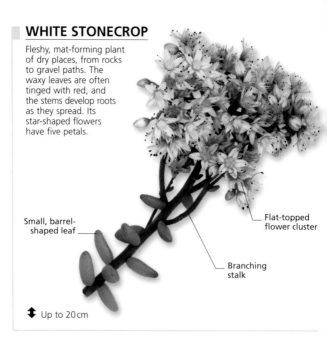

Small, barrel-shaped leaf

Flat-topped flower cluster

Branching stalk

✦ Up to 20 cm

MEADOWSWEET

Patch-forming plant of ditches and damp ground. From mid- to late summer, it has foam-like masses of fragrant flowers carried on stiff, upright stems. Its toothed leaflets feel rough to the touch.

Creamy white flower

Slim, rigid stem

Rounded bud

Toothed leaflet

✦ Up to 1.2 m

COW PARSLEY

Tall plant of grassy places and roadsides, where it often forms large colonies. Its flowers are arranged in umbrella-shaped flowerheads with up to 15 main spokes, which appear in early spring. Its fern-like leaves are attached to hollow stems.

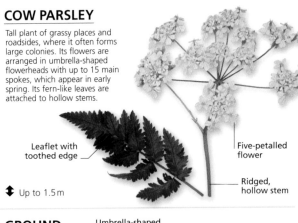

Leaflet with toothed edge

Five-petalled flower

Ridged, hollow stem

⬍ Up to 1.5 m

GROUND ELDER

Umbrella-shaped flowerhead

Clump-forming plant of shaded places, roadsides, and gardens, where it can be a troublesome weed. It spreads by underground stems. Its flowerheads have up to 20 main spokes, which grow on branching stalks.

Hollow stalk

Toothed leaflet

Cluster of five-petalled flowers

⬍ Up to 90 cm

WILD CARROT

Variable plant of field edges and dry grassland, particularly near coasts. Its finely divided leaves grow on ridged, branching stems. Its flowers form flat-topped umbrellas that close up as they ripen.

Spreading ruff of bracts

Ripening seedhead

Wiry stem

Finely divided leaf

⬍ Up to 1 m

»

CLEAVERS

Straggling, weak-stemmed plant of waste ground and hedgerows. Also known as Goosegrass. It climbs using tiny hooks on its stems and leaves. Its flowers produce hooked seedheads. These fall off when rubbed, latching on to clothing and fur.

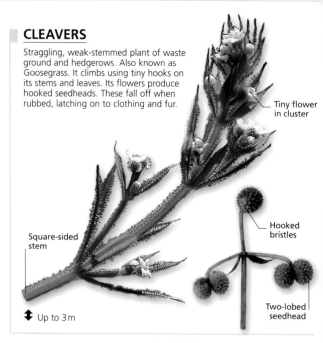

Tiny flower in cluster

Square-sided stem

Hooked bristles

✦ Up to 3 m

Two-lobed seedhead

YARROW

Patch-forming plant of dry grassland, roadsides, and embankments. Its deep green leaves are finely divided, which makes them look feathery. It spreads by means of underground stems. The small flowerheads range from off-white to pale pink in colour.

Unstalked upper leaf

Finely divided leaf

Flat-topped flowerhead

✦ Up to 80 cm

Stiff stem

LILY OF THE VALLEY

Low, patch-forming plant found in dry woodland and mountain grassland, and also as a garden plant. It spreads by means of underground stems. It has fragrant, drooping flowers in one-sided clusters. These are sometimes hidden by paired, elliptical leaves.

Dark green leaf

Bell-shaped flower

Erect stem

⬍ Up to 25 cm

»

Medium-sized Flowers

Medium-sized white flowers are common from spring to autumn. Some of them – such as Bramble and Wild Strawberry – produce edible fruits.

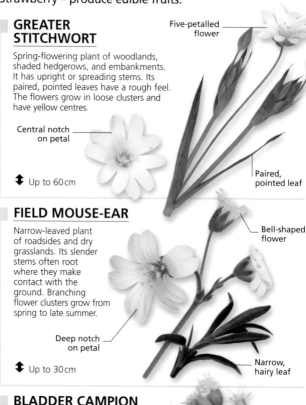

GREATER STITCHWORT

Spring-flowering plant of woodlands, shaded hedgerows, and embankments. It has upright or spreading stems. Its paired, pointed leaves have a rough feel. The flowers grow in loose clusters and have yellow centres.

Five-petalled flower

Central notch on petal

Paired, pointed leaf

✤ Up to 60 cm

FIELD MOUSE-EAR

Narrow-leaved plant of roadsides and dry grasslands. Its slender stems often root where they make contact with the ground. Branching flower clusters grow from spring to late summer.

Bell-shaped flower

Deep notch on petal

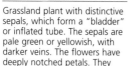

Narrow, hairy leaf

✤ Up to 30 cm

BLADDER CAMPION

Grassland plant with distinctive sepals, which form a "bladder" or inflated tube. The sepals are pale green or yellowish, with darker veins. The flowers have deeply notched petals. They open wide at night.

Pale green sepal

Oval, pointed leaf

Greyish stem

Petal with slender lobes

✤ Up to 80 cm

TRAVELLER'S JOY

Woody-stemmed climber typically found on lime-rich soil. It uses leaf stalks to cling to nearby supports. Its creamy white flowers produce silky-haired seedheads, which remain after the leaves fall. Also known as Old Man's Beard.

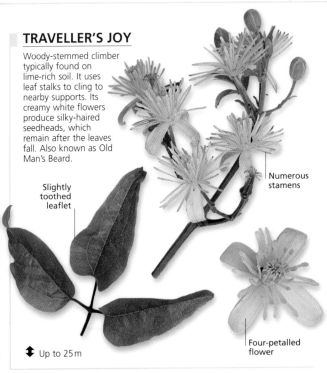

Numerous stamens

Slightly toothed leaflet

Four-petalled flower

✢ Up to 25 m

MEADOW SAXIFRAGE

Found in grassy places, this plant forms low rosettes, with kidney-shaped leaves close to the ground. In late spring and early summer, groups of 4–12 flowers grow at the tips of upright stems.

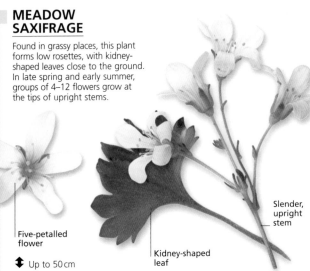

Slender, upright stem

Five-petalled flower

Kidney-shaped leaf

✢ Up to 50 cm

»

BRAMBLE

Prickly plant that grows in woods, hedges, and waste ground. Its long, arching stems root at their tips, creating extensive clumps. Its flowers are white, pink, or purple. They produce blackberries, which become soft and juicy when ripe.

Five-petalled flower

Backward-pointing prickle

‡ Up to 3 m

WILD RASPBERRY

Clump-forming shrub of woodlands and embankments, with upright, prickly stems. Its pale green leaves have white undersides. Its inconspicuous flowers produce eye-catching – and edible – bright red fruit.

Green sepal

Five-petalled flower

Toothed leaflet

‡ Up to 1.5 m

WILD STRAWBERRY

Five-petalled flower

A miniature relative of the cultivated strawberry, found in woodland and hedgerows, particularly on chalky soil. It has three-part leaves and bright red fruit. It spreads by runners that produce new plants. Its flowers are white and yellow.

Toothed leaf margin

Pea-sized fruit

‡ Up to 25 cm

WOOD SORREL

Patch-forming woodland plant. Five-petalled flowers are lined with pink veins and grow singly. Each leaf has three heart-shaped leaflets, which fold up at night and during rain. After flowering, the plant produces slender seedheads that burst apart if touched.

Bell-shaped flower

Leaflet folds along midrib

Pink vein

Heart-shaped leaflet

⬍ Up to 10 cm

HOGWEED

Robust wayside plant with coarse, divided leaves. Its umbrella-shaped flowerheads are up to 15 cm wide, with pinkish or white flowers. Petals are larger around outer rim. Its larger relative, Giant Hogweed, is poisonous and should not be touched.

Flat-topped flowerhead

Hollow, bristly stem

⬍ Up to 2.5 m

WHITE DEADNETTLE

Widespread plant of woodlands, hedgerows, and fertile disturbed ground. Its square-sided stems carry pairs of nettle-like leaves that do not sting. Its flowers grow in tightly packed whorls towards the tips of stems.

Toothed leaf

Two-lipped flower

⬍ Up to 50 cm

»

Daisy Family

With nearly 25,000 species worldwide, daisies and their relatives form one of the largest plant families. They include many wayside and grassland wildflowers.

The plants in this family have tiny flowers, known as florets. These are packed together in composite flowerheads, which are easy to mistake for single flowers. Each floret produces just one seed – look out for the hairs or hooks that help them to spread.

Winning formula

The Dandelion is one of the most successful daisy family plants, spreading far and wide with its windblown seeds. As well as dandelions and daisies, the family also includes ragworts, thistles, and knapweeds, and cultivated plants such as sunflowers, chicory, and lettuces.

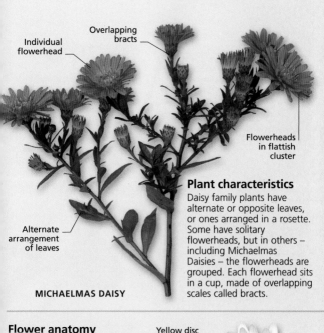

Individual
flowerhead

Overlapping
bracts

Flowerheads
in flattish
cluster

Alternate
arrangement
of leaves

MICHAELMAS DAISY

Plant characteristics

Daisy family plants have
alternate or opposite leaves,
or ones arranged in a rosette.
Some have solitary
flowerheads, but in others –
including Michaelmas
Daisies – the flowerheads are
grouped. Each flowerhead sits
in a cup, made of overlapping
scales called bracts.

Flower anatomy

A daisy flowerhead contains two
kinds of florets. Disc florets, at the
centre, are tubular, with five petals
shaped like tiny teeth. Ray florets,
at the outside, are also tubular, but
they have a large, petal-like flap.
Dandelion flowerheads have ray
florets only, while thistles have
only disc florets.

Yellow disc
florets

White ray florets

DAISY

Ray florets

Hollow
stem

DANDELION

Disc florets

Short,
spiny bract

SPEAR THISTLE

DAISY

Familiar plant of lawns and grassy places. Its stems are velvety. The leaves are crowded into basal rosettes. The low-growing flowerheads open in the morning, and close at dusk. The plant blooms year-round, except when very cold.

Unbranched stem

Yellow central disc

Spoon-shaped leaf

Petal-like ray tinged with pink

⬍ Up to 10 cm

WILD GARLIC

Patch-forming plant of damp, shady places, where it may carpet the ground. It has two or three leaves, and a strong garlic-like smell. Its flowers grow in rounded clusters, at the end of ridged stalks. Also known as Ramsons.

Star-shaped flower

Leafless flower stalk

Leaf-like spathes enclose flower buds

Elliptical leaf

⬍ Up to 45 cm

Large Flowers

Plants with large white flowers are easy to spot in hedges and waysides. One of them – the White Waterlily – grows in lakes and ponds.

WHITE WATERLILY

Instantly recognizable freshwater plant, with circular, dark green floating leaves. Its flowers are up to 20 cm wide, on stalks that emerge just above the water's surface. Wild plants are white, but there are many coloured garden forms.

Numerous stamens

Green-backed outer sepal

Circular leaf

⬍ In water up to 3 m

WOOD ANEMONE

Delicate, patch-forming woodland plant. Its nodding flowers bloom in early spring. They open wide in bright sunshine and are often tinged with pink underneath. Leaves grow with the flowers, and more appear with the seedheads.

Solitary flower

Deeply lobed leaf

⬍ Up to 30 cm

GREAT BINDWEED

Fast-growing climber often found in gardens and wasteland, sometimes smothering fences and other supports. Its twining stems carry alternate leaves. Usually all-white, the flowers sometimes have pink stripes.

Green pouch at base of flower

Trumpet-shaped flower

Heart-shaped leaf

⬍ Up to 3 m

»

FIELD ROSE

Scrambling rose found in hedgerows and scrub. Either free-standing or climbing through other plants. Its styles form a short column at the centre of each flower. The stems have scattered thorns. Its leaves are divided into five or seven leaflets. After flowering, it produces bright red rosehips.

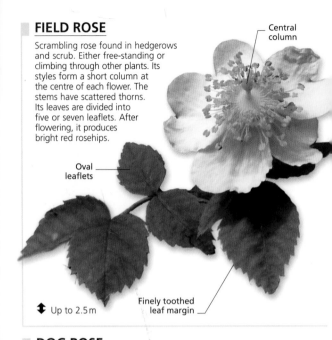

Central column

Oval leaflets

Finely toothed leaf margin

↕ Up to 2.5m

DOG ROSE

Clambering or free-standing rose that grows over hedges and in rough, grassy ground. Its flowers appear in early summer, and vary in colour between pale pink and white. Its arching stems are thorny and are often covered with red rosehips in autumn.

Pink-flushed petal

Yellow stamens

Spreading sepal

Toothed leaflet

↕ Up to 2.5m

HONEYSUCKLE

Climber of woods and hedgerows. Its blue-grey leaves are untoothed. The tubular flowers are strongly perfumed, particularly after dusk. They turn yellow with age and are followed by a cluster of red berries.

Curved lip

Protruding stamen

Paired oval leaves

Woody stem

✤ Up to 6 m

OXEYE DAISY

Conspicuous grassland plant of embankments, meadows, and roadsides. It stands out against surrounding grass. Much taller than a true daisy, it has stiff, branched stems, and white and yellow flowerheads up to 5 cm wide.

Spoon-shaped leaf

Yellow central disc

✤ Up to 80 cm

SNOWDROP

Clump-forming woodland plant that flowers in late winter and early spring. It has two slender leaves and a single green and white flower. It has three white sepals that are larger than its petals. The leaves persist after the flower has produced seeds.

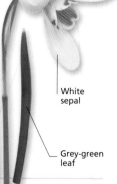

White sepal

Green and white petal

Nodding flower

Grey-green leaf

✤ Up to 20 cm

»

3 YELLOW FLOWERS

Wherever they grow, plants with yellow flowers are hard to miss. Their vibrant colour attracts a wide range of insect pollinators. Their flowering period often extends for weeks, although individual blooms may last for just a few days.

Small Flowers

Despite their small size, yellow flowers can be very conspicuous when they grow in clusters or flowerheads.

Tiny flower

WILD MIGNONETTE

Branching, bushy plant of disturbed ground and grasslands, usually on lime-rich soil. It has narrow, deeply divided leaves and smooth stems. Tall spires of flowers appear from mid- to late summer.

Six-petalled flower

Curled leaflet margin

↕ Up to 80 cm

AGRIMONY

Grassland plant with toothed, divided leaves and a tall column of bright yellow flowers. After flowering, it produces small, hooked seedheads. These latch onto clothes or animal fur, helping the plant to spread.

Flower bud

Tiny leaflet between larger leaflets

Five-petalled flower

↕ Up to 1 m

FENNEL

Tall plant of waste ground and dry, rocky places, often near the sea. It has umbrella-shaped flowerheads with up to 30 spokes, and highly divided leaves. The plant smells of aniseed if bruised or crushed.

Flat-topped cluster of tiny flowers

Bright yellow petal

Thread-like leaf

↕ Up to 2.5 m

RIBBED MELILOT

Smooth-stemmed, upright or straggling plant common on roadsides, grassy places, and bare ground. The slender stems are hairless, and the leaves are divided into threes. The tiny, yellow peaflowers are borne in one-sided clusters and produce brown, single-seeded pods.

Flowers hang downwards

Scalloped leaflet

⬍ Up to 1.5m

BLACK MEDICK

Low-growing, trailing plant common in grassland and disturbed ground. Its tiny flowers grow in rounded clusters, each about the size of a pea. The seed pods turn black when ripe.

Flower cluster

Long stalk

Leaf with three oval leaflets

Black seed pod

⬍ Up to 50cm

»

LADY'S BEDSTRAW

Upright or spreading plant of grassland, roadsides, and sand dunes. It has a large number of tiny, four-petalled flowers, which are fragrant. Its leaves are narrow and dark green, and grow in whorls along its stems.

Dense, branched flower cluster

Linear, shiny leaf

✤ Up to 80 cm

CANADIAN GOLDENROD

Clump-forming plant common throughout Europe in abandoned allotments and on waste ground. In late summer, its tall, upright stems are topped by flowerheads arranged in horizontal rows.

Golden yellow flowerhead

Narrow, toothed leaf

✤ Up to 2 m

Medium-sized Flowers

Plants with medium-sized yellow flowers include species in the pea family, and also relatives of daisies, which have flowerheads made of small florets.

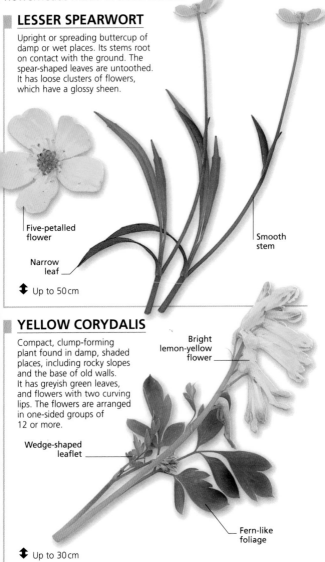

LESSER SPEARWORT

Upright or spreading buttercup of damp or wet places. Its stems root on contact with the ground. The spear-shaped leaves are untoothed. It has loose clusters of flowers, which have a glossy sheen.

Five-petalled flower

Narrow leaf

Smooth stem

✤ Up to 50 cm

YELLOW CORYDALIS

Compact, clump-forming plant found in damp, shaded places, including rocky slopes and the base of old walls. It has greyish green leaves, and flowers with two curving lips. The flowers are arranged in one-sided groups of 12 or more.

Bright lemon-yellow flower

Wedge-shaped leaflet

Fern-like foliage

✤ Up to 30 cm

»

WALLFLOWER

Originally from rocky places, including cliffs and old walls, this plant is also grown in gardens, where its fragrant flowers range from brick-red to mauve and white. In the wild form, they are golden yellow.

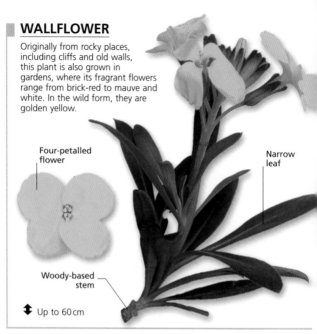

Four-petalled flower

Narrow leaf

Woody-based stem

✥ Up to 60 cm

WILD RADISH

Bristly wasteland plant, with lobed lower leaves and open clusters of flowers. Its flowers are yellow, white, or mauve, with a tracery of darker veins. The seeds develop inside pods, which have narrow joints and a pointed "beak".

Four-petalled flower

Narrow, veined petal

✥ Up to 75 cm

WALLPEPPER

Fleshy, low-growing plant of dry, rocky places, stone roofs, and old walls. It forms small yellowish green clumps, with clusters of star-shaped flowers in spring and early summer. It is also known as Biting Stonecrop.

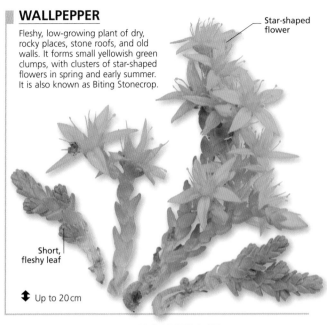

Star-shaped flower

Short, fleshy leaf

✤ Up to 20 cm

YELLOW MOUNTAIN SAXIFRAGE

Low-growing plant of damp mountain rocks and streamsides, at altitudes of up to 3,000 m. Its spreading stems form cushion-like clumps. It has clusters of flowers, with flowering and non-flowering shoots.

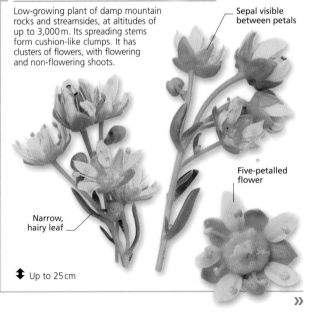

Sepal visible between petals

Five-petalled flower

Narrow, hairy leaf

✤ Up to 25 cm

»

CREEPING CINQUEFOIL

Low-growing plant of bare ground, waysides, and gardens, spreading by slender runners. Its leaves have 5–7 leaflets. Its flowers grow on long stalks. It can be a stubborn weed in cultivated ground.

Five-petalled flower

Toothed leaflet

Red-tinged stem

Hand-like leaf

Rounded petal with central notch

↕ Up to 10 cm

HERB BENNET

Slender, upright plant found in woodland and shaded places. Its flowers grow singly and they produce hooked seedheads, which cling to clothes and fur.

Rounded petal

Long stalk

Lobed stem leaf

Hooked seedhead

⬍ Up to 70 cm

SILVERWEED

Patch-forming plant of bare ground, grassy places, and roadsides. It spreads by overground runners or creeping stems. Its leaves are green or grey above, and silvery below. They have up to 25 leaflets each.

Silvery underside of leaf

Rounded petal

Cup-shaped flower

⬍ Up to 20 cm

»

Mint Family

Found worldwide, the mint family includes about 7,000 species. Apart from mints, it contains many wildflowers of open places or shaded ground.

Mints and their relatives have a long history of cultivation, with many being planted as culinary herbs. Wild plants are often easy to spot – look out for telltale whorls of two-lipped flowers, which grow on square, upright stems.

Spreading overground
Many plants in the mint family form dense mats or clumps. Some spread by underground stems, but Bugle (above) spreads by growing runners. New plants sprout at intervals where the runners straggle across the ground.

Plant characteristics

Most members of the mint family have square stems, and leaves arranged in opposite pairs. Their flowers often grow in closely packed whorls around their stems, with the lowest flowers opening first.

Tight whorl of flowers

Coarsely toothed leaf margin

Leaves in opposite pairs

Square stem

WHITE DEADNETTLE

Flower anatomy

Mint family flowers have a wide range of colours, from white to pink, yellow, and blue. They are typically two-lipped. The upper lip often forms a hood, while the lower lip makes up a landing platform for visiting insects. Each flower usually produces four seeds.

Upper lip

Lower lip

Sepals form funnel-shaped tube

YELLOW ARCHANGEL

GORSE

Spiny shrub found in heaths, grassland, and near coasts. Its coconut-scented flowers are tightly grouped. They appear throughout the year, peaking in late winter and spring. Mature stems lack leaves. The seeds develop in brown pods.

Golden yellow peaflower

Furrowed spine

↕ Up to 2 m

DYER'S GREENWEED

Spreading or upright miniature shrub found in grassy places, road verges, and open scrub. Its numerous stems are densely packed with clusters of peaflowers. Its small leaves are untoothed and narrow. Its seed pods are brown.

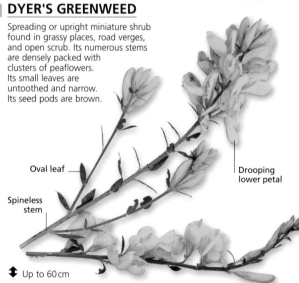

Oval leaf

Drooping lower petal

Spineless stem

↕ Up to 60 cm

MEADOW VETCHLING

Clambering plant of grassland and
roadsides. Clusters of 5–12 flowers
grow at the tips of slightly winged
stems. The divided leaves have
clinging tendrils and arrow-shaped
stipules at the base of the leaf stalks.
The seeds develop in black pods.

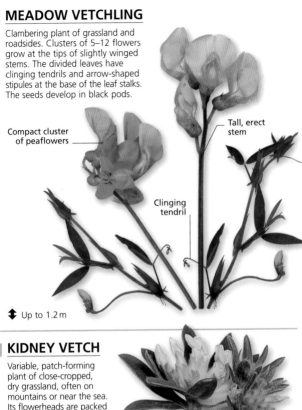

**Compact cluster
of peaflowers**

**Tall, erect
stem**

**Clinging
tendril**

✸ Up to 1.2 m

KIDNEY VETCH

Variable, patch-forming
plant of close-cropped,
dry grassland, often on
mountains or near the sea.
Its flowerheads are packed
with silky hairs. Tightly
grouped peaflowers may be
yellow, red, orange, or white.
Its leaves have up to 15 leaflets.

**Rounded
flowerhead**

**Ladder-shaped
leaf**

✸ Up to 50 cm

»

PERFORATE ST JOHN'S-WORT

Upright plant of dry fields and
rough grasslands. It spreads by
creeping and rooting at its base.
Its leaves have tiny, translucent
dots. Its five-petalled flowers
are dark yellow. These grow
in loose clusters on the
end of ridged stems.

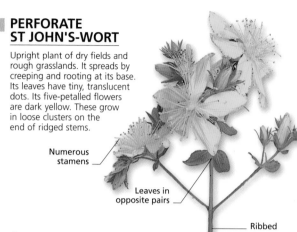

Numerous
stamens

Leaves in
opposite pairs

Ribbed
stem

✦ Up to 80 cm

COWSLIP

Spring-flowering plant of meadows
and grassy verges, sometimes
forming profuse colonies. Its nodding
flowers grow on upright stems.
Up to 30 nodding flowers grow in
one-sided clusters on each upright
stem. The bright green wrinkly
leaves form rosettes.

Five-petalled
flower

Orange
markings

Strap-like
leaf stalk

Spoon-shaped
leaf

✦ Up to 25 cm

YELLOW ARCHANGEL

Patch-forming woodland plant with
upright, square-sided stems. Its stalked
leaves are large and nettle-like. Its
bright yellow flowers are two-lipped
and grow in whorls. The plant has a
strong smell if crushed or bruised.

Hooded
upper lip

Lower lip
with red streaks

Hairy stem

✦ Up to 50 cm

COMMON FLEABANE

Grey, woolly plant found in wet meadows and ditches, often in extensive clumps. Its wrinkled, unstalked upper leaves clasp the stem. Its flowerheads have numerous, short, petal-like rays.

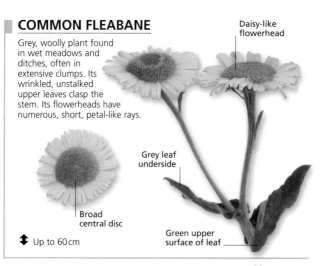

Daisy-like flowerhead

Grey leaf underside

Broad central disc

Green upper surface of leaf

✢ Up to 60 cm

COMMON RAGWORT

This weed of overgrazed or abandoned ground is poisonous to livestock. Its daisy-like flowerheads grow in large, flat-topped clusters on branching stems. In bloom from midsummer until autumn, the flowers produce seeds that are carried away by the wind.

Leaf with deeply divided lobes

Golden yellow flowerhead

Ridged, cylindrical stem

✢ Up to 1.5 m

PRICKLY OXTONGUE

A wayside plant common in grassy places and disturbed ground. Its pimply leaves have wavy edges and bristles. Its bright yellow flowerheads are streaked with red on the underside.

Petal-like ray

Branched stem

Long, narrow leaf

✢ Up to 80 cm

»

Large Flowers

From early spring onwards, large yellow flowers adorn many habitats, from woodlands and hedges to lakes and ponds.

▍YELLOW WATERLILY

Eye-catching plant of lakes and slow-flowing water, with oval floating leaves. Each leaf has a deep cleft where its stalk meets the blade. Its flowers are held just above the surface of the water.

Oval leaf

Curved stamens

Rounded flower

Flask-shaped seedhead

Deep cleft

✦ In water up to 5 m deep

▍MARSH MARIGOLD

Robust plant of marshland and wet woodland. It has upright or spreading stems and glossy leaves. Its brilliant yellow flowers appear in early spring. Also known as the Kingcup.

Saucer-shaped flower

Numerous stamens

Kidney-shaped leaf

✦ Up to 60 cm

CREEPING BUTTERCUP

Common in damp fields and grassy places, this plant spreads by surface runners, which can make it a persistent garden weed. Its flowers appear from spring to late summer. Its leaves are divided into three leaflets.

Numerous stamens

Glossy yellow flower

Wedge-shaped leaflet

⬍ Up to 50 cm

LESSER CELANDINE

Buttercup of damp places and broadleaved woodland. Its leaves are shiny and deep green. The flowers have many narrow petals and three sepals each. They appear early in spring and they open wide in bright sunshine.

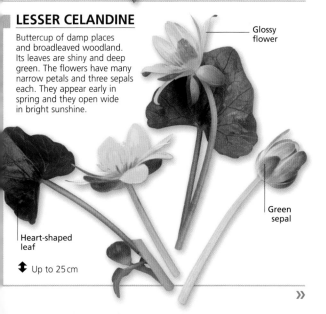

Glossy flower

Heart-shaped leaf

Green sepal

⬍ Up to 25 cm

»

WELSH POPPY

Clump-forming poppy
of damp, shaded places,
including rocky ground
and the base of old
walls. Its yellow or
orange-yellow flowers
grow singly on
slender stems.

Pale
green leaf

Four-petalled
flower

Toothed
lobe

Numerous
stamens

✦ Up to 60 cm

YELLOW HORNED POPPY

Robust, mound-forming
poppy with bright
yellow flowers. Its long
seed pods split open
when ripe. Found mainly
on coastal sand and shingle,
and rarely inland. It exudes
yellow sap if stems are cut.

Waxy petal

Fleshy
grey-green leaf

✦ Up to 90 cm

GREATER CELANDINE

Tall, buttercup-like member
of the poppy family, with
divided leaves and small
clusters of flowers.
Found in shady places,
including hedgerows
and the base of old
walls. Oozes orange
sap if stems are cut.

Pale
green leaf

Four-petalled
flower

Slender,
brittle stem

✦ Up to 90 cm

WILD CABBAGE

Thickset coastal plant with fleshy, undulating leaves and tall clusters of flowers. The base of its stem becomes woody with age. It is the wild ancestor of cabbage, broccoli, and many other cultivated vegetables.

Grey-green leaf

Thick midrib

Four-petalled flower

✸ Up to 2 m

BROOM

Densely branched shrub of heaths and grassy places, with green, flexible stems. It has fragrant peaflowers, grouped in open clusters, which produce flat-sided seed pods. The pods turn black when ripe.

Tiny leaf

Golden yellow flower

✸ Up to 2.5 m

COMMON ROCK-ROSE

Crumpled, papery petal

Small, ground-hugging plant with leaves that are dark green above and grey beneath. Found in grassland and rocky places, almost always on limestone or chalk. Its buds hang downwards, and each flower lasts for a single day.

Narrow leaf

Five-petalled flower

✸ Up to 50 cm

»

COMMON EVENING-PRIMROSE

Robust plant that grows on embankments, roadsides, and coastal sand. Covered in short hairs, the plant has tall stems and very large four-petalled flowers. Its buds open in the evening, and the bowl-shaped flowers wither the following day.

Stiff stem

Reddish sepal

Sharply pointed flower bud

⬍ Up to 1.5 m

PRIMROSE

Clump-forming plant of grassy verges and woodland. Its pale yellow flowers are a familiar sign of spring. The five-petalled flowers grow singly on slender stalks. The bright green leaves grow in a low rosette.

Corrugated upper surface

Notched petal

⬍ Up to 15 cm

Broad midrib

GREAT MULLEIN

Statuesque plant of wasteland and dry, stony ground. It is covered with soft, greyish hairs. The velvety leaves form a ground-hugging rosette in the first year. It produces a tall column of bright yellow flowers in its second year.

Densely packed flowerhead

Rounded petal

Five-petalled flower

Grey-green stem leaf

✤ Up to 2 m

COMMON TOADFLAX

Often seen on roadsides, this plant has linear, grey-green leaves. The flowers are two-lipped with a backward-pointing spur filled with nectar. The lower lip has orange markings. The flowers are pollinated by bumblebees, which push open the lips to reach the nectar inside.

Bright yellow flower

Large, oval seed case

Tapering spur

✤ Up to 80 cm

»

COLTSFOOT

Patch-forming plant of damp places
and roadsides. Its flowerheads grow
singly on scale-covered stems,
appearing well before the
leaves in early spring.
The black-edged leaves
have angled sides and
a downy underside.

Purplish
scale

Petal-like
ray

Toothed
leaf margin

↕ Up to 25 cm

COMMON HAWKWEED

Variable plant of grassy and rocky
places. Its main leaves are crowded
in a flat rosette near the ground.
The slender stems bear flat-topped
clusters of bright yellow flowerheads
in summer. The seeds later spread
in the wind.

Slender,
leafless
stem

Petal-like
ray

↕ Up to 80 cm

DANDELION

Widespread and variable
plant that grows in many
habitats, from urban parks to
meadows and scrub. It has long
leaves arranged in a basal rosette.
The broad flowerheads produce a
"clock" of seeds with attached
"parachutes". The plant
oozes a milky juice if cut.

Jagged
leaf

Hollow
flowering
stem

Petal-like
ray floret

↕ Up to 40 cm

WILD DAFFODIL

Ancestor of cultivated daffodils. Found in broadleaved woodland and old meadows, where it can form extensive clumps. Its flowers have deep yellow trumpets and six paler tepals, or outer lobes. Cultivated forms, which are generally larger, also sometimes grow in the wild.

Yellow stamen

Triangular tepal

Grey-green strap-like leaf

⬍ Up to 50 cm

YELLOW FLAG

Robust, clump-forming iris that grows in marshes, ponds, riverbanks, and the margins of canals. Its leaves are pointed and light green. It has eye-catching flowers with three upright "standard" petals, and three broader "fall" petals, which are streaked with dark veins.

Fall petal

Flat, spear-like leaf

⬍ Up to 1.5 m

4 RED–PINK FLOWERS

Pure red is a rare colour in Europe's wildflowers. Pink, on the other hand, is much more common and widespread. Pink flowers sometimes grade into purple, so look in Chapter 5 (pp.88–103) if you cannot find a pink flower here.

Small Flowers

Small red or pink flowers sometimes grow singly, but in most plants they are grouped together in clusters or flowerheads.

COMMON BISTORT

Clump-forming plant of damp, grassy places and old pastures, where it spreads through underground stems. Its leaves are untoothed, with the highest ones clasping the flowering stems. Dense flowerheads appear from early summer to autumn.

Untoothed leaf margin

Five-petalled flower

Cylindrical flowerhead

⬍ Up to 1 m

SHEEP'S SORREL

Common plant of dry meadows, where its flowerheads can form large patches in the grass. Its arrow-shaped leaves have a mildly acidic taste. The tiny flowers grow in branched clusters, with male and female flowers on different plants.

Rusty red flowerhead

Hairless stem

Leaf with three sharp points

⬍ Up to 30 cm

MOSS CAMPION

Ground-hugging mountain plant with a moss-like shape that protects it from cold winds. The slender leaves are tightly packed. Its five-petalled flowers grow singly on short stalks, creating a mound of colour when in bloom.

Protruding stamens

Short flower stalk

Bright green leaf

✤ Up to 10 cm

GREATER SAND-SPURREY

Mat-forming plant of coasts and salt marshes, and occasionally salty soil inland. It has whorls of cylindrical leaves, with papery sheaths at their base. The flowers have short, green sepals and pale pink flowers with yellow anthers.

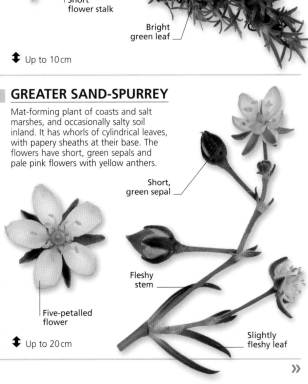

Short, green sepal

Fleshy stem

Five-petalled flower

✤ Up to 20 cm

Slightly fleshy leaf

»

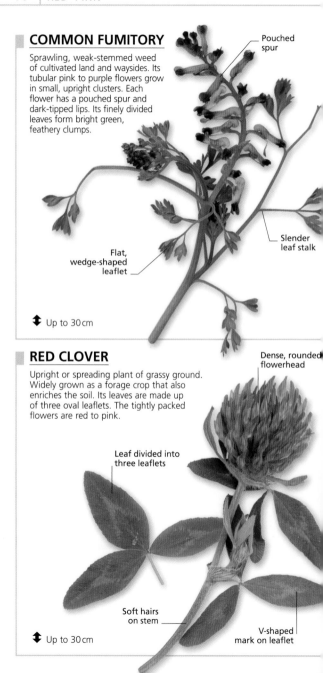

COMMON FUMITORY

Sprawling, weak-stemmed weed of cultivated land and waysides. Its tubular pink to purple flowers grow in small, upright clusters. Each flower has a pouched spur and dark-tipped lips. Its finely divided leaves form bright green, feathery clumps.

Pouched spur

Flat, wedge-shaped leaflet

Slender leaf stalk

✤ Up to 30 cm

RED CLOVER

Upright or spreading plant of grassy ground. Widely grown as a forage crop that also enriches the soil. Its leaves are made up of three oval leaflets. The tightly packed flowers are red to pink.

Dense, rounded flowerhead

Leaf divided into three leaflets

Soft hairs on stem

V-shaped mark on leaflet

✤ Up to 30 cm

HEATHER

Small evergreen shrub found on moors, heaths, and bogs. This plant grows slowly, sometimes carpeting the ground. Its thickly branched stems are covered with tiny leaves in opposite pairs. From midsummer to autumn, they bear branching clusters of small flowers.

Scale-like leaf

Woody stem

Pale pink to purple flower

✤ Up to 80 cm

BILBERRY

Dwarf deciduous shrub of heaths, moors, and woodland. Its tiny flowers are green or muddy red. The stems are ridged and angled. Its alternate leaves are bright green. After flowering, the plant produces edible, blue-black fruit.

Toothed leaf margin

✤ Up to 50 cm

Bell-shaped flower

»

SEA ROCKET

Found on sand and shingle beaches, this light green plant sometimes grows close to the high-tide mark. It has fleshy stems and leaves. Its pink or white flowers grow in tight clusters, with the stem lengthening as the seeds form.

Four-petalled flower

Bullet-shaped seed capsule

Leaf usually deeply lobed

Hairless stem

↕ Up to 50 cm

THRIFT

Mound-forming plant found on coastal cliffs and rocks, and mountains inland. Its leaves persist all year round. Its compact, rounded flowerheads grow on upright stalks.

Rounded flowerhead

Papery scale

Slender stalk

Tough, grass-like leaf

↕ Up to 30 cm

COMMON SEA-LAVENDER

Common late-flowering plant of saltmarshes, where it often carpets the ground. Its leaves are elliptical or spoon-shaped, and are attached at ground level. The flowers grow in spreading clusters and keep their colour if they are cut and dried.

Lilac-pink flower

Single main vein on leaf

↕ Up to 50 cm

SCARLET PIMPERNEL

Low-growing plant often found in farmland. It has weak, four-cornered stems. Its reddish orange flowers have five petals. They open wide in bright weather and close when it rains.

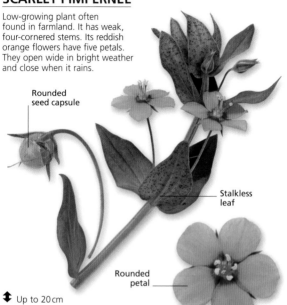

Rounded
seed capsule

Stalkless
leaf

Rounded
petal

✦ Up to 20 cm

WILD THYME

Low-growing, mat-forming shrub found in chalk grassland and stony places. Its flowers grow in small rounded heads and are often pollinated by butterflies. It has an aromatic smell if its leaves are brushed or bruised.

Protruding
style and
stamens

Small,
oval leaf

Pinkish
purple flower

✦ Up to 10 cm

»

HEMP AGRIMONY

Tall, clump-forming plant common in damp, shaded ground. It has paired leaves with toothed leaflets. Blooming in late summer, its tiny, tubular, pink flowers grow in domed heads. They are rich in nectar and often teem with bees and butterflies.

Protruding stamens

Robust, dark red stem

 Up to 1.5m

SLENDER THISTLE

One of many thistles with pink flowerheads, this tall and very prickly plant has alternate leaves that are lobed and spiny. Its flowerheads grow on side branches, as well as at the top of the plant.

Crowded, unstalked flowerhead

Prickly "wing" on stem

Spiny leaf

 Up to 75cm

PYRAMIDAL ORCHID

Late-flowering orchid found in chalk downland and other grassy places. It is named after its flowerheads, which have a triangular shape when they first appear. The pink, purple, or white flowers are pollinated by butterflies and moths. The lance-shaped leaves are pale green and unspotted.

Three-lobed lip

Backward-pointing spur on flower

Up to 60cm

Medium-sized Flowers

Medium-sized red or pink flowers occur in many plants, from pinks themselves to deadnettles and orchids. They include some common weeds.

MAIDEN PINK

Tuft-forming plant of dry, sandy grassland and rocky slopes. Its flowers bloom from midsummer to early autumn. They are deep pink, with a ring-like marking at the centre. Its bluish green leaves grow in opposite pairs.

Ragged petal edge

Ring-like marking

Slender leaf

✦ Up to 30 cm

CUCKOO FLOWER

Spring-flowering plant that grows in damp meadows and shaded verges. Its pink or lilac four-petalled flowers grow in loose clusters at the tips of the stems.

Rounded, overlapping petal

Narrow upper leaf

✦ Up to 60 cm

ORPINE

Tall, fleshy plant found in woodland edges and grassy places, on dry, sandy soil. It blooms in mid- to late summer, with buds maturing slowly to form domed clusters of star-shaped flowers. Its alternate leaves are fleshy and pale green.

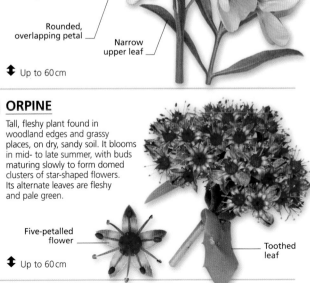

Five-petalled flower

Toothed leaf

✦ Up to 60 cm

»

Cress Family

Europe has many wildflowers from the cress family, out of a total of 3,500 species worldwide. All of them have an easy-to-spot feature: four petals, arranged in a cross.

Cress family plants, or crucifers, grow in a variety of habitats, from farmland and waysides to rocky coasts and cliffs. They include some common weeds, such as Shepherd's Purse, as well as the wild ancestors of cabbages, mustard, and other crops.

Facing the sea

Sea Rocket (above) is one of several cress family plants that grow close to the shore. It often grows on sand dunes, where its fleshy leaves resist salt-laden spray, as well as strong coastal winds.

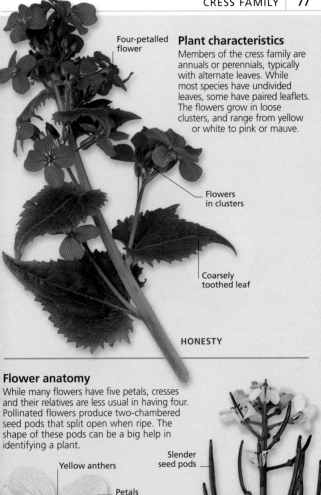

Four-petalled flower

Plant characteristics

Members of the cress family are annuals or perennials, typically with alternate leaves. While most species have undivided leaves, some have paired leaflets. The flowers grow in loose clusters, and range from yellow or white to pink or mauve.

Flowers in clusters

Coarsely toothed leaf

HONESTY

Flower anatomy

While many flowers have five petals, cresses and their relatives are less usual in having four. Pollinated flowers produce two-chambered seed pods that split open when ripe. The shape of these pods can be a big help in identifying a plant.

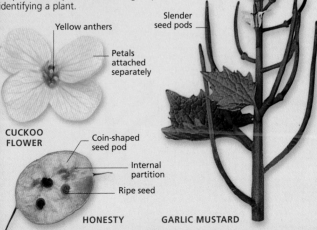

Yellow anthers

Petals attached separately

CUCKOO FLOWER

Slender seed pods

Coin-shaped seed pod

Internal partition

Ripe seed

HONESTY

GARLIC MUSTARD

COMMON RESTHARROW

Creeping, clump-forming plant found in meadows and grassland, particularly on chalky soil. Its leaves are divided into three oval leaflets. The flowers are pink or purplish, with petals arranged in typical pea family shape (see pp.96–97). They produce hairy seed pods, each containing one or two seeds.

Five-petalled flower

Leaflet with toothed edge

Stem becomes woody with age

↕ Up to 50 cm

CROWN VETCH

This grassland plant is instantly recognizable by its bicoloured flowerheads. It has sprawling stems and divided, ladder-like leaves. Its flowers are arranged in elegant domes or "crowns" at the top of leafless stalks.

"Crown" of lilac and white flowers

Elliptical leaflet

Slender stem

↕ Up to 1 m

HERB ROBERT

Common weed of gardens and shady places. Its flowers usually grow in pairs. The plant is often tinged with red, and has a distinctive smell if touched or crushed.

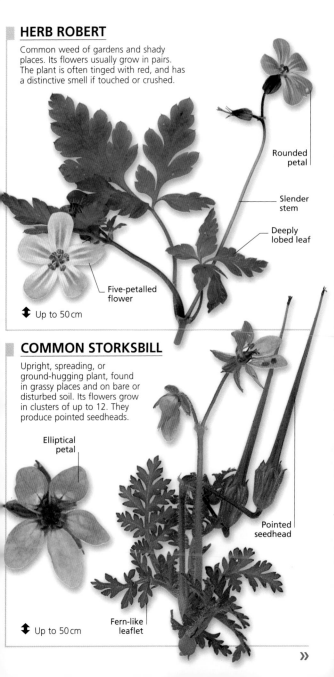

Rounded petal

Slender stem

Deeply lobed leaf

Five-petalled flower

↕ Up to 50 cm

COMMON STORKSBILL

Upright, spreading, or ground-hugging plant, found in grassy places and on bare or disturbed soil. Its flowers grow in clusters of up to 12. They produce pointed seedheads.

Elliptical petal

Pointed seedhead

Fern-like leaflet

↕ Up to 50 cm

»

COMMON COMFREY

Robust, clump-forming plant from waysides and damp places. Its upper leaves are joined to the stem through a pair of "wings". Pink, purplish, or white flowers grow in coiled clusters, with the end of each coil flowering last.

Spear-shaped leaf

Untoothed leaf margin

Funnel-shaped flower

✦ Up to 1.5m

RED DEADNETTLE

Common garden weed that is unrelated to true nettles and does not sting. Its low-growing stems are topped with soft, bristly leaves. Its unusually long flowering season lasts from March to December.

Two-lipped flower

Purplish upper leaf

Coarsely toothed leaf

Square-sided stem

✦ Up to 25cm

RED VALERIAN

Robust plant from dry, rocky places; often seen sprouting from old walls. Its grey-green leaves are borne in opposite pairs. The flowers are pink, red, or white, and produce seeds that are carried away on feathery "parachutes".

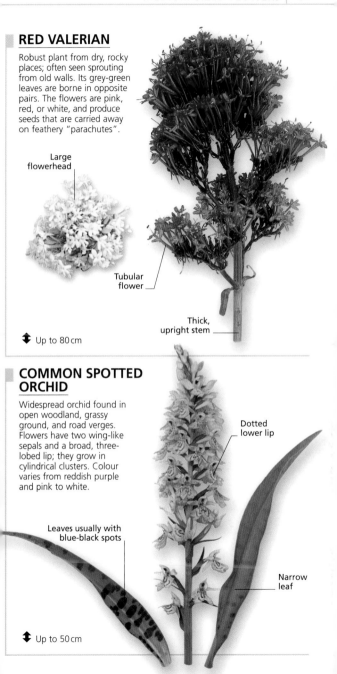

Large flowerhead

Tubular flower

Thick, upright stem

⬍ Up to 80 cm

COMMON SPOTTED ORCHID

Widespread orchid found in open woodland, grassy ground, and road verges. Flowers have two wing-like sepals and a broad, three-lobed lip; they grow in cylindrical clusters. Colour varies from reddish purple and pink to white.

Dotted lower lip

Leaves usually with blue-black spots

Narrow leaf

⬍ Up to 50 cm

Large Flowers

This group of flowers includes some of Europe's most colourful wildflowers, such as Red Campion, the Common Poppy, and the unmistakable Foxglove.

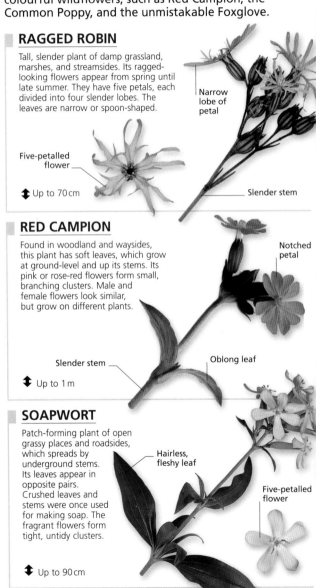

RAGGED ROBIN

Tall, slender plant of damp grassland, marshes, and streamsides. Its ragged-looking flowers appear from spring until late summer. They have five petals, each divided into four slender lobes. The leaves are narrow or spoon-shaped.

Narrow lobe of petal

Five-petalled flower

↨ Up to 70 cm

Slender stem

RED CAMPION

Found in woodland and waysides, this plant has soft leaves, which grow at ground-level and up its stems. Its pink or rose-red flowers form small, branching clusters. Male and female flowers look similar, but grow on different plants.

Notched petal

Slender stem

Oblong leaf

↨ Up to 1 m

SOAPWORT

Patch-forming plant of open grassy places and roadsides, which spreads by underground stems. Its leaves appear in opposite pairs. Crushed leaves and stems were once used for making soap. The fragrant flowers form tight, untidy clusters.

Hairless, fleshy leaf

Five-petalled flower

↨ Up to 90 cm

COMMON POPPY

One of Europe's most unmistakable wildflowers and a common farmland plant. The four-petalled flowers may be up to 10 cm wide and usually only last for a single day.

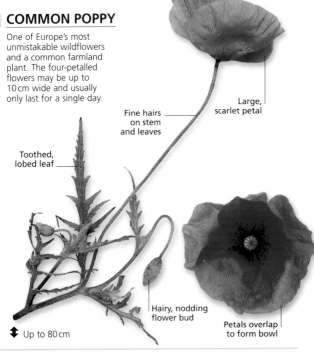

Large, scarlet petal

Fine hairs on stem and leaves

Toothed, lobed leaf

Hairy, nodding flower bud

Petals overlap to form bowl

Up to 80 cm

HIMALAYAN BALSAM

Introduced from Asia, this very tall, fast-growing plant is now well established in damp ground along Europe's rivers and streams. Its stems have whorls of leaves and pouched flowers. The seed capsules explode when ripe.

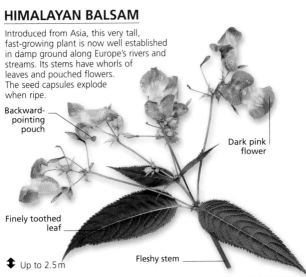

Backward-pointing pouch

Dark pink flower

Finely toothed leaf

Fleshy stem

Up to 2.5 m

»

COMMON MALLOW

Widespread plant of waste
ground and grassy places.
It blooms from mid- to
late summer. The pink
or purple flowers have
well-separated petals
and grow in clusters. Its
sprawling or upright stems
grow from a deep taproot.

Rounded,
softly hairy
leaf

Five-petalled
flower with
dark veins

Central notch
on petal

⬍ Up to 1 m

TREE MALLOW

Woody-based plant of coastal rocks
and shingle. It is covered with soft,
felty hairs. The leaves are hand-shaped,
with five to seven lobes. The flowers
have pink to purple petals and a
central tuft of pink stamens.

Dark-veined
petal

Wavy lobe

Large,
cup-shaped
flower

Stout stem

⬍ Up to 2.5 m

ROSEBAY WILLOWHERB

Extremely vigorous, patch-forming plant of waste ground and waysides. It spreads by underground stems. The tall, upright stems bear closely packed leaves. Its rose-pink flowers grow in tall spires. After flowering, they release many fluffy seeds.

Flower bud

Slender leaf with pale midrib

Prominent stamens

↕ Up to 1.5m

BINDWEED

Disliked by gardeners, this plant excels at spreading with its creeping, underground stems. Above ground, it twines around other plants, or spreads out over bare ground. Its flowers are pink, white, or a stripy mixture of the two colours.

Trumpet-shaped flower

Stalked, arrow-shaped leaf

Yellow centre

↕ Up to 1.5m

FOXGLOVE

Handsome plant of woodland, heaths and waysides. First-year plants form a rosette of softly hairy leaves. From their second year, plants produce one-sided spires of flowers, whose petals are joined to form a tube.

Downward-hanging, spotted flower

Many-seeded fruit capsule

↕ Up to 2m

»

SPEAR THISTLE

Common in waste places, disturbed ground, and farmland, where it can be a troublesome weed. This large, prickly plant has reddish purple flowerheads with a narrow "waist", flaring out above into a brush-like shape.

Narrow waist

Spear-shaped, spiny leaf

Spiny wing along stem

✦ Up to 1.5m

COMMON KNAPWEED

Thistle-like plant of meadows and road verges. Unlike thistles, its long, pointed leaves have no spines. The flowerheads grow singly or in branched clusters. Their florets sit in a dark brown cup of overlapping scales called bracts.

Long, pointed leaf

Brown bract

Reddish to purple florets

Slender, rigid stem

✦ Up to 1m

BEE ORCHID

Low-growing orchid of meadows and embankments. It starts growing in autumn and blooms the following spring. The flower resembles and even smells like a female bee – a mimicry that attracts pollinating male wasps and bees.

Leaf edge turns inwards

Slender stem

Lowest petal forms furry lip

Three pink sepals

✦ Up to 45 cm

EARLY PURPLE ORCHID

Early-flowering orchid of open woodland, scrub, and road verges. Its pink or purple flowers grow in a narrow column and have a long, broad spur. Its spreading sepals are folded backwards.

Upward-pointing spur

Fleshy stem

Spotted, dark green leaf

✦ Up to 50 cm

5 PURPLE–BLUE FLOWERS

Some of Europe's most eye-catching plants have purple or blue flowers. They include some well-known spring favourites, such as forget-me-nots and bluebells, as well as common flowers of late summer, which bring the season to a close.

Small Flowers

Flowers in this group are commonly found in grassy places and along roadsides. They vary in colour from sky-blue to violet and purple.

TUFTED VETCH

Clambering plant of grassy places and roadsides, where it can be spotted by its abundant violet-blue pea-like flowers. It clings to surrounding vegetation by slender tendrils at the end of its leaves.

Slender tendril

One-sided flower cluster

Compact flower buds

Hairy stem

Paired leaflet

↕ Up to 1.5 m

COMMON MILKWORT

A delicate plant, usually seen in short turf on chalk or limestone. It has upright and spreading stems that carry clusters of blue, pink, or white flowers. The lower of the three small petals is fringed, and the larger sepals form a colourful pair of wings.

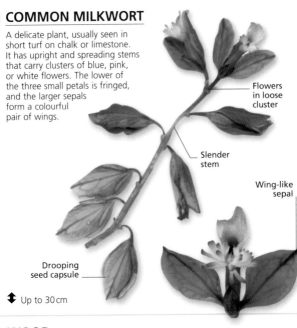

Flowers in loose cluster

Slender stem

Wing-like sepal

Drooping seed capsule

✤ Up to 30 cm

WOOD FORGET-ME-NOT

One of many kinds of forget-me-not, this plant is found in woodland clearings and damp meadows. It is covered with soft hairs, and has small, sky-blue flowers with flat, rounded petals and yellow throats.

Coiled cluster of flower buds

Five-petalled flower

Soft hairs

Oval, untoothed leaf

✤ Up to 50 cm

»

BUGLE

Square-stemmed, patch-forming plant common on woodland edges and damp, shady ground. It has opposite pairs of oval leaves and bracts, which are often tinged with copper or metallic blue. Its flowers are blue, or sometimes pink or white.

Leafy bract

Prominent lower lip

Oval leaf

Square stem

✤ Up to 30 cm

COMMON FIELD-SPEEDWELL

Low-growing plant of fields and disturbed ground, where it blooms almost year-round. Borne on slender stalks, the flowers open wide in sunshine, but close quickly if the sky becomes overcast.

Toothed leaf margin

Four-petalled flower

Lowermost petal often pale

Oval, pale green leaf

✤ Up to 20 cm

Medium-sized Flowers

Plants with medium-sized purple or blue flowers range from ground-hugging species to climbers or clamberers that use other plants for support.

COMMON VETCH

Clambering plant of cultivated ground, roadsides, and waste places. Its divided leaves end in forked tendrils. The flowers grow singly or in pairs and, after withering, produce seeds in slender pods.

Forked tendrils

Sharp-tipped leaflet

Two-toned flower

Ridged stem

✤ Up to 1.5 m

SWEET VIOLET

Low-growing plant of woodlands, hedges, and roadsides. It produces drifts of sweet-smelling flowers in early spring. The flowers are violet or white, with contrasting colours sometimes growing on neighbouring plants.

Solitary flower

Slender stalk

Kidney-shaped leaf

White variant

✤ Up to 15 cm

PURPLE LOOSESTRIFE

Found on riverbanks, in ditches, and other wet habitats, this clump-forming plant has square-sided stems, and leaves arranged in twos or threes. The flowers grow in tall flowerheads, with the lowest buds opening first.

Bright purple flower

Stalkless leaf

Ridged stem

✤ Up to 1.5 m

»

SPRING GENTIAN

Mat-forming plant of damp mountain meadows, where it blooms soon after the snow melts in spring. It has opposite pairs of ground-hugging leaves, and each bright blue flower has a deep central tube, coloured white.

Spoon-shaped petal

Oval leaf

Five-petalled flower

⬍ Up to 7.5 cm

VIPER'S BUGLOSS

Tall, eye-catching plant of disturbed ground and dry open places. It forms a ground-hugging rosette of bristly leaves in its first year. Flowering stems appear in the second year, covered with purple-blue blooms that are highly attractive to bees.

Protruding pink stamens

Narrow leaf

Funnel-shaped flower

⬍ Up to 1 m

GROUND IVY

Low-growing plant of damp shaded soil, where it can form extensive mats, sometimes making it a troublesome weed. Its soft, velvety leaves grow on long stalks. Its violet-blue or pink flowers grow in whorls up its stems.

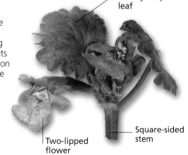

Kidney-shaped leaf

Square-sided stem

Two-lipped flower

⬍ Up to 25 cm

BITTERSWEET

Scrambling plant of scrub and hedges, easily recognized by its purple and yellow flowers. It has large leaves with 3–5 lobes. Clusters of flowers produce oval berries, initially green but red when ripe. The whole plant is poisonous.

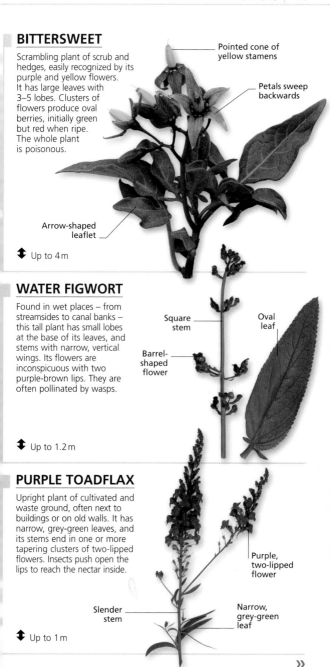

Pointed cone of yellow stamens

Petals sweep backwards

Arrow-shaped leaflet

✦ Up to 4 m

WATER FIGWORT

Found in wet places – from streamsides to canal banks – this tall plant has small lobes at the base of its leaves, and stems with narrow, vertical wings. Its flowers are inconspicuous with two purple-brown lips. They are often pollinated by wasps.

Square stem

Oval leaf

Barrel-shaped flower

✦ Up to 1.2 m

PURPLE TOADFLAX

Upright plant of cultivated and waste ground, often next to buildings or on old walls. It has narrow, grey-green leaves, and its stems end in one or more tapering clusters of two-lipped flowers. Insects push open the lips to reach the nectar inside.

Purple, two-lipped flower

Slender stem

Narrow, grey-green leaf

✦ Up to 1 m

»

Pea Family

With their unique shape, pea family flowers are some of the easiest to recognize. There are nearly 20,000 species across the world, with over 800 in Europe alone.

Pea family plants, or legumes, grow in many different habitats, but they are particularly common in grassland, along hedgerows, and around the edges of fields. In Europe, there are some woody species – including Gorse and Broom – but far more are soft-stemmed climbers, which use other plants for support.

Yellow parade

Gorse (above) has a very long flowering season, reaching its peak in late winter and early spring. It has typical peaflowers, which produce hard seed pods. The pods snap open in warm weather, scattering their seeds far from the parent plant.

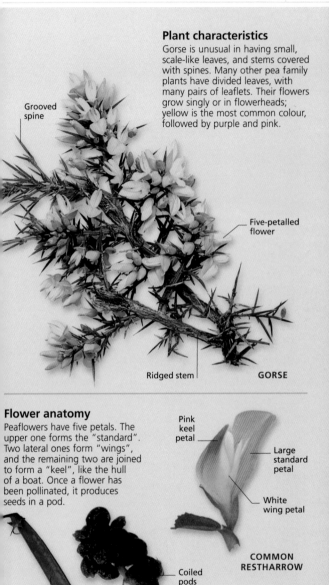

Plant characteristics

Gorse is unusual in having small, scale-like leaves, and stems covered with spines. Many other pea family plants have divided leaves, with many pairs of leaflets. Their flowers grow singly or in flowerheads; yellow is the most common colour, followed by purple and pink.

Grooved spine

Five-petalled flower

Ridged stem

GORSE

Flower anatomy

Peaflowers have five petals. The upper one forms the "standard". Two lateral ones form "wings", and the remaining two are joined to form a "keel", like the hull of a boat. Once a flower has been pollinated, it produces seeds in a pod.

Pink keel petal

Large standard petal

White wing petal

COMMON RESTHARROW

Straight seed pod

Coiled pods

BROAD-LEAVED EVERLASTING PEA

BLACK MEDICK

IVY-LEAVED TOADFLAX

Often found in cracks in old walls, this plant has trailing stems and ivy-shaped leaves. Its lilac or mauve flowers grow on slender stalks. After flowering, the stalks bend away from light, dropping the seeds in crevices, where they germinate.

Fleshy, lobed leaf

Purple, thread-like stem

Yellow patch on lower lip

✦ Up to 25 cm

HAREBELL

Delicate, slender plant of meadows, roadsides, and grassy places, as well as rocky ground and dunes. Its leaves are rounded at ground level and narrow on the stems. In mid- and late summer, it produces open clusters of nodding flowers.

Bell-shaped flower

Narrow leaf

Thread-like stem

✦ Up to 50 cm

BLUEBELL

Carpeting the ground in deciduous woodland, this plant blooms as the trees come into leaf. The leaves emerge at ground level, withering by the time the plant produces seeds. Flowers grow in clusters, with up to 15 hanging to one side on each stalk.

Smooth, leafless stalk

Bell-shaped flower

Dark green leaf

✦ Up to 50 cm

Large Flowers

Some large purple or blue flowers appear in early spring. Others – such as Michaelmas Daisy and Chicory – bloom as summer comes to an end.

MONKSHOOD

Robust, poisonous plant from meadows, damp woodland, and the banks of streams. Its stem is erect and unbranched. The large leaves are divided into branching lobes, and blue or violet flowers appear in tall clusters. The top of each flower forms a conspicuous hood, which gives the plant its name.

✦ Up to 1.5 m

Closely packed flower cluster

Deeply divided leaf

PASQUE-FLOWER

Found in dry, grassy places – sometimes on mountains – this low-growing plant has eye-catching flowers and feathery leaves. The flowers have purple sepals with contrasting yellow stamens. They grow on velvety stems with a collar of leafy bracts.

Finely divided leaf

✦ Up to 30 cm

Leafy bract

Petal-like sepal

COLUMBINE

Attractive but poisonous plant of damp meadows and grassy roadsides, usually growing in lime-rich soil. Its nodding flowers have hooked spurs on their petals. Wild plants are usually blue, but garden varieties with different colours can also be found.

Hooked spur

Rounded leaflet

✦ Up to 90 cm

Upright, flowering stem

»

HONESTY

Robust plant of waste ground
and waysides, where its flowers
appear in the middle of spring.
It has large, deep green leaves,
and purple or white flowers
in dense clusters. After
flowering, it produces flat,
oval seed cases that persist
after the plant has died.

Four-petalled
flower

Bluntly toothed
leaf margin

Heart-shaped
leaf

✤ Up to 1 m

BROAD-LEAVED
EVERLASTING PEA

Vigorous, climbing or
sprawling plant from grassy
places, roadsides, and scrub. The
purple-pink or occasionally white
flowers appear from mid- to late
summer. Its stems have two broad
wings. The leaves are divided
into two leaflets and have
long branching tendrils.

Compact
flower
cluster

Pair of
leaflets

Purplish
pink
petal

✤ Up to 3 m

MEADOW CRANESBILL

Conspicuous plant of meadows
and road verges, particularly on
limestone or chalky soils. Its
flowers grow in loose clusters,
held above the foliage. The
leaves are cut into
branching segments.

Lighter vein
on petal

Deeply
divided leaf

Violet-blue,
five-petalled
flower

✤ Up to 1 m

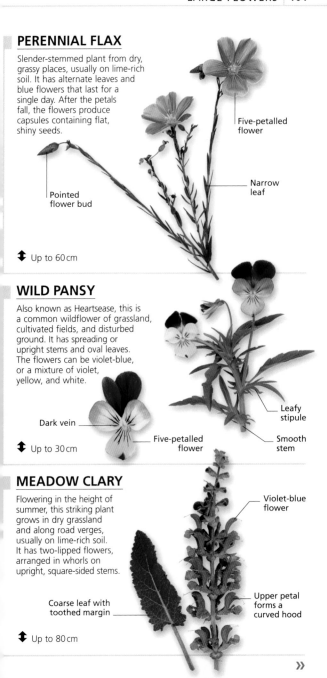

PERENNIAL FLAX

Slender-stemmed plant from dry, grassy places, usually on lime-rich soil. It has alternate leaves and blue flowers that last for a single day. After the petals fall, the flowers produce capsules containing flat, shiny seeds.

Five-petalled flower

Narrow leaf

Pointed flower bud

↕ Up to 60 cm

WILD PANSY

Also known as Heartsease, this is a common wildflower of grassland, cultivated fields, and disturbed ground. It has spreading or upright stems and oval leaves. The flowers can be violet-blue, or a mixture of violet, yellow, and white.

Leafy stipule

Dark vein

Five-petalled flower

Smooth stem

↕ Up to 30 cm

MEADOW CLARY

Flowering in the height of summer, this striking plant grows in dry grassland and along road verges, usually on lime-rich soil. It has two-lipped flowers, arranged in whorls on upright, square-sided stems.

Violet-blue flower

Coarse leaf with toothed margin

Upper petal forms a curved hood

↕ Up to 80 cm

»

TEASEL

Tall, prickly plant of grassy places and embankments, easily recognized by its size, its bristly flowerheads, and its unusual leaves. The leaves grow in pairs and meet at their bases, creating a hollow that often fills with rain.

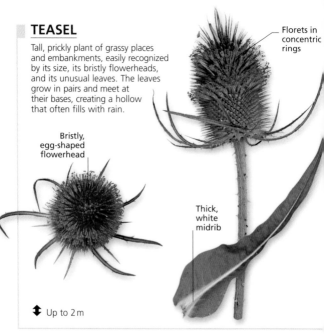

Florets in concentric rings

Bristly, egg-shaped flowerhead

Thick, white midrib

✿ Up to 2 m

DEVIL'S-BIT SCABIOUS

Widespread plant of dry or damp grassy places, where it flowers from midsummer to autumn. It has oval leaves clustered near the ground, and narrower leaves on its stems. The flowerheads contain many small flowers, or florets.

Protruding stamens

Budding flowerhead

Button-shaped flowerhead

Prominent midrib on leaf

✿ Up to 1 m

MICHAELMAS DAISY

Originally from North America, this tall, clump-forming plant grows on waste ground, roadsides, and stream banks, where it flowers in late summer and autumn. Its flowerheads have purple or blue to white flowers.

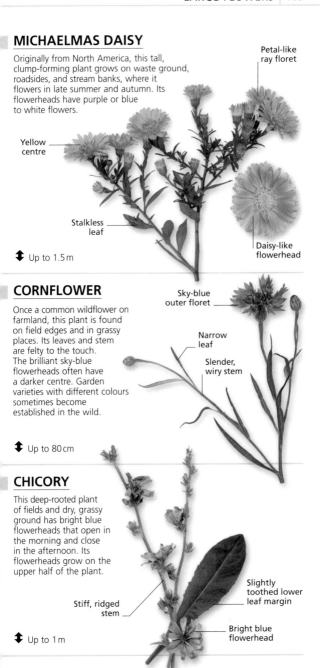

Petal-like ray floret

Yellow centre

Stalkless leaf

Daisy-like flowerhead

✤ Up to 1.5 m

CORNFLOWER

Once a common wildflower on farmland, this plant is found on field edges and in grassy places. Its leaves and stem are felty to the touch. The brilliant sky-blue flowerheads often have a darker centre. Garden varieties with different colours sometimes become established in the wild.

Sky-blue outer floret

Narrow leaf

Slender, wiry stem

✤ Up to 80 cm

CHICORY

This deep-rooted plant of fields and dry, grassy ground has bright blue flowerheads that open in the morning and close in the afternoon. Its flowerheads grow on the upper half of the plant.

Stiff, ridged stem

Slightly toothed lower leaf margin

Bright blue flowerhead

✤ Up to 1 m

FLOWER GALLERY

This gallery shows the wildflowers already profiled in the book, grouped by family. A family consists of closely related genera, which are, in turn, made up of related species that often look similar. You can use this gallery if you think you know the family the flower is in or to find out which one it belongs to. Then go to its profile page to learn all about it.

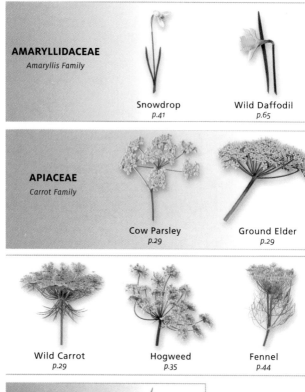

AMARYLLIDACEAE
Amaryllis Family

Snowdrop
p.41

Wild Daffodil
p.65

APIACEAE
Carrot Family

Cow Parsley
p.29

Ground Elder
p.29

Wild Carrot
p.29

Hogweed
p.35

Fennel
p.44

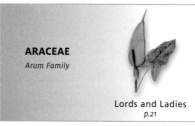

ARACEAE
Arum Family

Lords and Ladies
p.21

ASTERACEAE
Daisy Family

Yarrow
p.30

Daisy
p.38

Oxeye Daisy
p.41

Canadian Goldenrod
p.46

Common Fleabane
p.57

Common Ragwort
p.57

Prickly Oxtongue
p.57

Coltsfoot
p.64

Common Hawkweed
p.64

Dandelion
p.64

Hemp Agrimony
p.74

Slender Thistle
p.74

Spear Thistle
p.86

Common Knapweed
p.86

Michaelmas Daisy
p.103

»

Cornflower
p.103

Chicory
p.103

BALSAMINACEAE
Balsam Family

Himalayan Balsam
p.83

BORAGINACEAE
Forget-me-not Family

Common Comfrey
p.80

Wood Forget-me-not
p.91

Viper's Bugloss
p.94

BRASSICACEAE
Cress Family

Garlic Mustard
p.26

Sweet Alison
p.26

Shepherd's Purse
p.27

Watercress
p.27

Wallflower
p.48

Wild Radish
p.48

Wild Cabbage
p.61

Sea Rocket
p.72

Cuckoo Flower
p.75

Honesty
p.100

CAMPANULACEAE
Bellflower Family

Harebell
p.98

CAPRIFOLIACEAE
Honeysuckle Family

Honeysuckle
p.41

CARYOPHYLLACEAE
Pink Family

Greater Stitchwort
p.32

Field Mouse-ear
p.32

Bladder Campion
p.32

Moss Campion
p.69

Greater Sand-spurrey
p.69

Maiden Pink
p.75

»

»

Ragged Robin
p.82

Red Campion
p.82

Soapwort
p.82

CHENOPODIACEAE
Goosefoot Family

Sea Beet
p.19

Fat Hen
p.19

Glasswort
p.19

CISTACEAE
Rock-rose Family

Common Rock-rose
p.61

CONVOLVULACEAE
Bindweed Family

Great Bindweed
p.39

Bindweed
p.85

CRASSULACEAE
Stonecrop Family

Navelwort
p.20

White Stonecrop
p.28

Wallpepper
p.49

Orpine
p.75

DIPSACACEAE
Teasel Family

Teasel
p.102

Devil's-bit Scabious
p.102

ERICACEAE
Heather Family

Heather
p.71

Bilberry
p.71

EUPHORBIACEAE
Spurge Family

Petty Spurge
p.20

FABACEAE
Pea Family

Ribbed Melilot
p.45

Black Medick
p.45

»

»

Gorse
p.54

Dyer's Greenweed
p.54

Meadow Vetchling
p.55

Kidney Vetch
p.55

Broom
p.61

Red Clover
p.70

Common Restharrow
p.78

Crown Vetch
p.78

Tufted Vetch
p.90

Common Vetch
p.93

Broad-leaved
Everlasting Pea
p.100

GENTIANACEAE
Gentian Family

Spring Gentian
p.94

GERANIACEAE
Geranium Family

Herb Robert
p.79

Common Storksbill
p.79

Meadow Cranesbill
p.100

HYPERICACEAE
Hypericum Family

Perforate
St John's-wort
p.56

IRIDACEAE
Iris Family

Yellow Flag
p.65

LAMIACEAE
Mint Family

White Deadnettle
p.35

Yellow Archangel
p.56

Wild Thyme
p.73

Red Deadnettle
p.80

Bugle
p.92

Ground Ivy
p.94

Meadow Clary
p.101

LILIACEAE
Lily Family

Lily of the Valley
p.31

Wild Garlic
p.38

»

»

Bluebell
p.98

LINACEAE
Flax Family

Perennial Flax
p.101

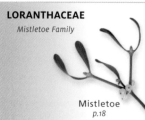

LORANTHACEAE
Mistletoe Family

Mistletoe
p.18

LYTHRACEAE
Loosestrife Family

Purple Loosestrife
p.93

MALVACEAE
Mallow Family

Common Mallow
p.84

Tree Mallow
p.84

NYMPHACEAE
Waterlily Family

White Waterliy
p.39

Yellow Waterlily
p.58

ONAGRACEAE
Willowherb Family

Common Evening-
Primrose
p.62

Rosebay Willowherb
p.85

ORCHIDACEAE
Orchid Family

Pyramidal Orchid
p.74

Common Spotted Orchid
p.81

Bee Orchid
p.87

Early Purple Orchid
p.87

OXALIDACEAE
Wood Sorrel Family

Wood Sorrel
p.35

PAPAVERACEAE
Poppy Family

Yellow Corydalis
p.47

Welsh Poppy
p.60

Yellow Horned Poppy
p.60

Greater Celandine
p.60

Common Fumitory
p.70

Common Poppy
p.83

PLANTAGINACEAE
Plantain Family

Ribwort Plantain
p.20

PLUMBAGINACEAE
Thrift Family

Thrift
p.72

»

Common Sea-lavender
p.72

POLYGALACEAE
Dock Milkwort Family

Common Milkwort
p.91

POLYGONACEAE
Smartweed Dock Family

Curled Dock
p.18

Common Bistort
p.68

Sheep's Sorrel
p.68

PRIMULACEAE
Primrose Family

Cowslip
p.56

Primrose
p.62

Scarlet Pimpernel
p.73

RANUNCULACEAE
Buttercup Family

Stinking Hellebore
p.21

Traveller's Joy
p.33

Wood Anemone
p.39

Lesser Spearwort
p.47

Marsh Marigold
p.58

Creeping Buttercup
p.59

Lesser Celandine
p.59

Monkshood
p.99

Pasque-flower
p.99

Columbine
p.99

RESEDACEAE
Mignonette Family

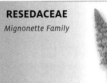

Wild Mignonette
p.44

ROSACEAE
Rose Family

Meadowsweet
p.28

Bramble
p34

Wild Raspberry
p.34

Wild Strawberry
p.34

Field Rose
p.40

Dog Rose
p.40

Agrimony
p.44

Creeping Cinquefoil
p.50

Herb Bennet
p.51

»

»

Silverweed
p.51

RUBIACEAE
Bedstraw Family

Cleavers
p.30

Lady's Bedstraw
p.46

SAXIFRAGACEAE
Saxifrage Family

Meadow Saxifrage
p.33

Yellow Mountain
Saxifrage
p.49

SCROPHULARIACEAE
Figwort Family

Great Mullein
p.63

Common Toadflax
p.63

Foxglove
p.85

Common
Field-Speedwell
p.92

Water Figwort
p.95

Purple Toadflax
p.95

Ivy-leaved Toadflax
p.98

SOLANACEAE
Potato Family

Bittersweet
p.95

URTICACEAE
Nettle Family

Stinging Nettle
p.18

VALERIANACEAE
Valerian Family

Red Valerian
p.81

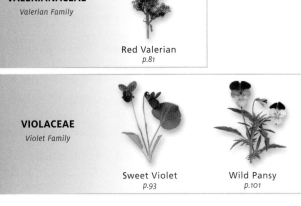

VIOLACEAE
Violet Family

Sweet Violet
p.93

Wild Pansy
p.101

Scientific Names

The scientific name of every species consists of two Latin words. The first one is the genus, which is common to closely related species that often look similar. The second is the specific name. The combination of these two words is unique to a particular species.

Common name	Scientific name	Page
Stinging Nettle	*Urtica dioica*	18
Mistletoe	*Viscum album*	18
Curled Dock	*Rumex crispus*	18
Sea Beet	*Beta vulgaris*	19
Fat Hen	*Chenopodium album*	19
Glasswort	*Salicornia europaea*	19
Navelwort	*Umbilicus rupestris*	20
Petty Spurge	*Euphorbia peplus*	20
Ribwort Plantain	*Plantago lanceolata*	20
Stinking Hellebore	*Helleborus foetidus*	21
Lords and Ladies	*Arum maculatum*	21
Garlic Mustard	*Alliaria petiolata*	26
Sweet Alison	*Lobularia maritima*	26
Shepherd's Purse	*Capsella bursa-pastoris*	27
Watercress	*Rorippa nasturtium-aquaticum*	27
White Stonecrop	*Sedum album*	28
Meadowsweet	*Filipendula ulmaria*	28
Cow Parsley	*Anthriscus sylvestris*	29
Ground Elder	*Aegopodium podagraria*	29
Wild Carrot	*Daucus carota*	29
Cleavers	*Galium aparine*	30
Yarrow	*Achillea millefolium*	30
Lily of the Valley	*Convallaria majalis*	31
Greater Stitchwort	*Stellaria holostea*	32
Field Mouse-ear	*Cerastium arvense*	32
Bladder Campion	*Silene vulgaris*	32
Traveller's Joy	*Clematis vitalba*	33
Meadow Saxifrage	*Saxifraga granulata*	33
Bramble	*Rubus fruticosus*	34
Wild Raspberry	*Rubus idaeus*	34
Wild Strawberry	*Fragaria vesca*	34
Wood Sorrel	*Oxalis acetosella*	35
Hogweed	*Heracleum sphondylium*	35
White Deadnettle	*Lamium album*	35
Daisy	*Bellis perennis*	38
Wild Garlic	*Allium ursinum*	38
White Waterlily	*Nymphaea alba*	39
Wood Anemone	*Anemone nemorosa*	39

Great Bindweed	*Calystegia sylvatica*	39
Field Rose	*Rosa arvensis*	40
Dog Rose	*Rosa canina*	40
Honeysuckle	*Lonicera periclymenum*	41
Oxeye Daisy	*Leucanthemum vulgare*	41
Snowdrop	*Galanthus nivalis*	41
Wild Mignonette	*Reseda lutea*	44
Agrimony	*Agrimonia eupatoria*	44
Fennel	*Foeniculum vulgare*	44
Ribbed Melilot	*Melilotus officinalis*	45
Black Medick	*Medicago lupulina*	45
Lady's Bedstraw	*Galium verum*	46
Canadian Goldenrod	*Solidago canadensis*	46
Lesser Spearwort	*Ranunculus flammula*	47
Yellow Corydalis	*Pseudofumaria lutea*	47
Wallflower	*Cheiranthus cheiri*	48
Wild Radish	*Raphanus raphanistrum*	48
Wallpepper	*Sedum acre*	49
Yellow Mountain Saxifrage	*Saxifraga aizoides*	49
Creeping Cinquefoil	*Potentilla reptans*	50
Herb Bennet	*Geum urbanum*	51
Silverweed	*Potentilla anserina*	51
Gorse	*Ulex europaeus*	54
Dyer's Greenweed	*Genista tinctoria*	54
Meadow Vetchling	*Lathyrus pratensis*	55
Kidney Vetch	*Anthyllis vulneraria*	55
Perforate St John's-wort	*Hypericum perforatum*	56
Cowslip	*Primula veris*	56
Yellow Archangel	*Lamiastrum galeobdolon*	56
Common Fleabane	*Pulicaria dysenterica*	57
Common Ragwort	*Senecio jacobaea*	57
Prickly Oxtongue	*Picris echioides*	57
Yellow Waterlily	*Nuphar lutea*	58
Marsh Marigold	*Caltha palustris*	58
Creeping Buttercup	*Ranunculus repens*	59
Lesser Celandine	*Ranunculus ficaria*	59
Welsh Poppy	*Meconopsis cambrica*	60
Yellow Horned Poppy	*Glaucium flavum*	60
Greater Celandine	*Chelidonium majus*	60
Wild Cabbage	*Brassica oleracea*	61
Broom	*Cytisus scoparius*	61
Common Rock-rose	*Helianthemum nummularium*	61
Common Evening-primrose	*Oenothera biennis*	62
Primrose	*Primula vulgaris*	62
Great Mullein	*Verbascum thapsus*	63
Common Toadflax	*Linaria vulgaris*	63
Coltsfoot	*Tussilago farfara*	64

Common Hawkweed	*Hieracium vulgatum*	64
Dandelion	*Taraxacum officinale*	64
Wild Daffodil	*Narcissus pseudonarcissus*	65
Yellow Flag	*Iris pseudacorus*	65
Common Bistort	*Persicaria bistorta*	68
Sheep's Sorrel	*Rumex acetosella*	68
Moss Campion	*Silene acaulis*	69
Greater Sand-spurrey	*Spergularia media*	69
Common Fumitory	*Fumaria officinalis*	70
Red Clover	*Trifolium pratense*	70
Heather	*Calluna vulgaris*	71
Bilberry	*Vaccinium myrtillus*	71
Sea Rocket	*Cakile maritima*	72
Thrift	*Armeria maritima*	72
Common Sea-lavender	*Limonium vulgare*	72
Scarlet Pimpernel	*Anagallis arvensis*	73
Wild Thyme	*Thymus polytrichus*	73
Hemp Agrimony	*Eupatorium cannabinum*	74
Slender Thistle	*Carduus tenuiflorus*	74
Pyramidal Orchid	*Anacamptis pyramidalis*	74
Maiden Pink	*Dianthus deltoides*	75
Cuckoo Flower	*Cardamine pratensis*	75
Orpine	*Sedum telephium*	75
Common Restharrow	*Ononis repens*	78
Crown Vetch	*Coronilla varia*	78
Herb Robert	*Geranium robertianum*	79
Common Storksbill	*Erodium cicutarium*	79
Common Comfrey	*Symphytum officinale*	80
Red Deadnettle	*Lamium purpureum*	80
Red Valerian	*Centranthus ruber*	81
Common Spotted Orchid	*Dactylorhiza fuchsii*	81
Ragged Robin	*Lychnis flos-cuculi*	82
Red Campion	*Silene dioica*	82
Soapwort	*Saponaria officinalis*	82
Common Poppy	*Papaver rhoeas*	83
Himalayan Balsam	*Impatiens glandulifera*	83
Common Mallow	*Malva sylvestris*	84
Tree Mallow	*Lavatera arborea*	84
Rosebay Willowherb	*Chamerion angustifolium*	85
Bindweed	*Convolvulus arvensis*	85
Foxglove	*Digitalis purpurea*	85
Spear Thistle	*Cirsium vulgare*	86
Common Knapweed	*Centaurea nigra*	86
Bee Orchid	*Ophrys apifera*	87
Early Purple Orchid	*Orchis mascula*	87
Tufted Vetch	*Vicia cracca*	90
Common Milkwort	*Polygala vulgaris*	91

Glossary

The terms defined here can help you to understand wildflowers better and to describe them with a greater degree of precision.

Alternate An arrangement where leaves grow singly at different levels along the stem.

Annual A plant that completes its life cycle in less than a year.

Anther The pollen-bearing part at the end of the stamen of a flower.

Bract A leaf-like organ at the base of a flower stalk.

Basal Located at the base of the plant.

Cultivated A plant specially bred or improved, or grown for its produce.

Disc floret The small, tubular flowers at the centre of the flowerhead of members of the daisy family.

Divided A leaf that is split into distinct leaflets or lobes.

Elliptical Describes a leaf with a wide middle and ends that taper to points.

Fall One of three outer petals that droop down in flowers of the iris family.

Family A scientific grouping of closely related plants.

Floret One of a group of small, individual flowers usually clustered together to form a flowerhead.

Flowerhead A cluster of florets or larger flowers.

Fruit The ripened dry or fleshy seed-bearing structure of a flowering plant.

Genus (pl. Genera) A category in classification consisting of a group of closely related species, and denoted by the first part of the scientific name, e.g. *Bellis* in *Bellis perennis*.

Keel The lower, fused petals of a peaflower.

Leaflet One of the leaf-like structures that make up a divided leaf.

Lip A protruding petal, as found in members of the orchid and mint families.

Lobe An often-rounded part of a divided leaf, formed by incisions towards the midrib.

Midrib The often-prominent central vein on a leaf or leaf-like structure.

Nectar A sugary fluid produced by some flowers, which draws insect pollinators.

Node The point from which leaves sprout on a stem.

Opposite Describes leaves growing in pairs along the stem.

Peaflower A flower, usually from the pea family, with sepals fused into a short tube and a distinctive petal arrangement.

Perennial A plant with a life cycle that spans several years.

Ray/Ray floret The small, flattened flowers that are usually found around the edge of the flowerhead in members of the daisy family.

Rhizome A horizontal fleshy stem, usually growing underground.

Rosette A circular cluster of leaves arranged at or near the base of a stem.

Runner A stem that creeps along the ground, forming roots at intervals and, eventually, separate plants.

Sap The watery fluid that carries nutrients and other substances around inside a plant.

Seedhead A flowerhead in seed.

Semi-parasitic A plant capable of photosynthesis that also derives some of its food from a host.

Sepal One of the separate, and usually green, parts of the flower that enclose the petals.

Spathe A large, hooded bract.

Spur A hollow, cylindrical, or pouched structure projecting from a flower, usually containing nectar.

Stamen The male reproductive organ of a flower, consisting of a stalk bearing an anther.

Standard The upright, upper petal of a peaflower, often larger than the others.

Stigma The part of the flower that receives pollen.

Stipule A leaf-like organ at the base of a leaf stalk.

Style The part of the female reproductive organ that joins the ovary to the pollen-receiving stigma.

Taproot A large, single root that grows vertically downwards, and from which other roots sprout.

Tendril A slender organ used by climbing plants to cling to a supporting object.

Tepal A part of the outside of a flower in which the petals and sepals are indistinguishable.

Toothed A leaf margin with indentations.

Umbel A flat-topped or domed cluster with stalks that rise from a common point.

Weed An invasive plant that grows where it is not wanted – disrupting the habitat and threatening the survival of existing plants, or affecting crops on farmland.

Whorl An arrangement where leaves grow in rings.

Wing One of the lateral petals of many flowers, particularly orchids or peaflowers.

Index

Page numbers in **bold** indicate main entry.